Cover photo
Collapse of the 40-year old I-35W bridge over the Mississippi river, Minneapolis, USA, 1 August 2007 (Tim Davis, EIT, www.conphoto.net)

Structures and Buildings

Proceedings of the
Institution of Civil Engineers
August 2008
Volume 161, issue SB4

GW01393019

Institution of Civil Engineers

CONTENTS

Structures and Buildings

Aims and scope

Structures and Buildings seeks the highest quality papers for publication. Papers on design and construction of civil engineering structures and buildings, and the associated applied research, will be considered. Papers could be on design aspects such as strength and durability of structural components, or the behaviour of buildings and bridge components and systems. Specific areas include M&E services, cladding, refurbishment, energy conservation, and people movement within/around buildings.

INSTRUCTIONS TO AUTHORS

Instructions to authors can be obtained online at www.structuresandbuildings.com or from www.editorialmanager.com/sb.

INTERNET AND E-MAIL CONTENTS ALERT

Access to all ICE and Thomas Telford journals is now available. For details of the full range of services, see www.thomastelford.com/journals or contact the Journals Department of Thomas Telford Ltd. Journal contents are available by e-mail as each issue is published. To register for this free service, go to the journal's website: www.structuresandbuildings.com

ICE VIRTUAL LIBRARY

The ICE Virtual Library contains more than 20 000 illustrated papers published in the ICE *Proceedings* since 1836. The papers are available for download at £15 each (special discount for ICE members).

SUBSCRIPTIONS

ICE members

Members receive discounts on all Proceedings journals. Please send enquiries and notification of changes of address to the Membership Registry, Institution of Civil Engineers, PO Box 4479, London SW1P 3XB.
Tel: +44 20 7 665 2227. Fax: +44 20 7 222 3514.
E-mail: subs@ice.org.uk
Claims for non-receipt of issues must be made within four months of publication or they will not be honoured without charge.

Libraries and non-members

Subscriptions to the individual journals comprising the Proceedings of the Institution of Civil Engineers are available at the individual prices shown in Table 1. A reduced rate package for all journals is also available. Please contact The Customer Services Department, Thomas Telford Ltd, 1 Heron Quay, London E14 4JD. Tel: +44 20 7 665 2460. Fax: +44 20 7 538 9620. E-mail: journals@thomastelford.com. Advertising enquiries should be directed to Warners Group

Origination by Keytec Typesetting Ltd, Bridport.
Printed in Great Britain by Bell & Bain Ltd, Glasgow.
ISSN 0965 0911 (Print) 1751-7702 (Online)
© The authors and the Institution of Civil Engineers, 2008

Publications, The Maltings, West Street, Bourne, Lincolnshire, PE10 9PH. Tel: +44 1778 391067.
E-mail: julietg@warnersgroup.co.uk

COPYRIGHT

DISCLAIMER

Papers or other contributions and the statements made or opinions expressed therein are published on the understanding that the author of the contribution is solely responsible for the opinions expressed in it and that its publication does not necessarily imply that such statements and/or opinions are or reflect the views or opinions of the ICE Council or ICE Committees.

The printing processes used to produce the *Proceedings of the Institution of Civil Engineers* minimise energy and water consumption. Waste paper, card, aluminium plates, ink tins and cartridges are recycled. Our printers are accredited to ISO 9001: 2000 quality standard. The journals are printed on chlorine-free paper, which meets the following environmental accreditations: ISO 14001 (production, emissions and use of energy); EMAS (Environmental Management System Certification); ISO 9706 (anti-ageing/ANSI standard).

Title	No. of issues	UK: £	Elsewhere by air: £
Bridge Engineering	4	216	237
Civil Engineering	6	123	142
Construction Materials	4	187	217
Energy	4	187	217
Engineering and Computational Mechanics	4	187	217
Engineering Sustainability	4	226	237
Geotechnical Engineering	6	180	209
Ground Improvement	4	217	254
Management, Procurement and Law	4	187	217
Maritime Engineering	4	158	174
Municipal Engineer	4	118	137
Structures and Buildings	6	303	353
Transport	4	145	167
Urban Design and Planning	4	187	217
Waste and Resource Management	4	187	217
Water Management	6	234	259
Package comprising all 16 titles	72	2426	2777

Table 1. Subscription prices for 2008

Proceedings of the Institution of
Civil Engineers
Structures & Buildings 161
August 2008 Issue SB4
Pages 169–170
doi: 10.1680/stbu.2008.161.4.169

Guest Editor
David Blockley, Emeritus
Professor and Senior
Research Fellow, University
of Bristol, UK

Editorial: Structural risk

D. Blockley PhD, DSc, FREng, FICE, FIStructE

A SPECIAL PUBLICATION OF THE INSTITUTION OF CIVIL ENGINEERS WITH THE INSTITUTION OF STRUCTURAL ENGINEERS

The purpose of this special issue is to present reviews of the latest research thinking about the risk, reliability and vulnerability of civil engineering structures. Some researchers argue that structural risk is a mature discipline. The personal view of the guest editor (and not necessarily of the guest contributors to this issue) is that, while it has made strong advances, much more development is required before it can smoothly be incorporated into practice. On the human scale, I estimate that it is about 16 years old: strong and growing up fast but still lacking maturity. All of the papers are aimed at nurturing that important development which will eventually ease the topic into mainstream practice. In that respect all of the papers are intended to be of interest to practitioners.

This issue has eight invited papers covering topics from the vulnerability of sensitive and historic structures in seismic zones to the risk issues behind the Eurocodes.

We start with the vulnerability of sensitive and historically important monumental structures. With its vast number of ancient churches, palaces and other antiquities, the problem is particularly acute in Italy. Theoretically simple risk registers are just inadequate. One reason is that the calculations do not prevent the 'double counting' of evidence. Bernardini and Lagomarsino describe some of the latest thinking in evaluating historic structures in seismic zones. A knowledge-based methodology based on the European macroseismic scale is used to develop criteria for the classification of buildings. It is based on evidence from differing sources varying from poor statistical data to detailed descriptions of single monuments. The authors use some of the latest interesting theoretical tools based on non-classical evidence theories such as fuzzy sets, random sets and imprecise probabilities, which avoid double counting. Vulnerability curves have been derived based on recent Italian earthquakes. The work shows that churches appear to be particularly vulnerable.

The effects of a relatively small explosion in a block of flats at Ronan Point in London in 1968 were disproportionate. The terrible tragedy of 9/11 and the collapse of the World Trade Center in New York have again exposed how theoreticians have neglected structural robustness. Agarwal and England review the literature. Structural engineering is mature, they

say, but it is often difficult to foresee where damage might occur that results in consequences that are disproportionate to the degree of damage. A robust structure will be able to cope with unexpected demands. There is, as yet, no agreed definition of robustness and hence no satisfactory measure. The difficulty is that we are dealing with risks that contain low chance/high consequence events. They point out that, while most studies assume a model of the loading conditions, robustness is also importantly a property of the form or connectedness of the structure.

Faced with the challenges of climate change and the need for sustainable design, structural risk is especially important for developing countries. Sánchez-Silva and Rosowsky state that developing countries need infrastructure development that is environmentally sound, socially acceptable, economically justifiable and sustainable. Financial pressures make long-term planning difficult. A balance between protecting the physical environment and using resources effectively is required. Engineers need models that are able to use and process large amounts of disparate information. As developing countries adopt the safety standards of developed countries then resources are directed away from other needs. Safety needs to be managed in the total context of the needs of the country.

Ellingwood is a world leader in the development of structural reliability theory. He recognises the early resistance to probabilistic methods but says that methods have developed to such a point that they are being used in modern codified design such as Eurocodes. In his paper he is concerned with how we deal with unforeseen events outside the traditional design envelope. He argues that infrastructure performance-based engineering offers a new paradigm. The important idea is that structural engineers must look beyond minimum code requirements if they are to meet the challenges ahead. Risks of unexpected events cannot be avoided. Risk-informed performance-based engineering requires a continuing dialogue among the project team and stakeholders with clear audit trails for key decisions about risk and safety. Uncertainty analysis must be a central part of the decision model. Tradeoffs must be treated candidly with a transparent decision process.

Vrouwenvelder summarises the treatment of risk and reliability in structural Eurocodes. He discusses how the Eurocodes deal with decision making under uncertainty. The links between partial factors and the underlying theory are highlighted by distinguishing between probability based design and fully probabilistic methods. He regrets that the theory seems to be understood by only a handful of specialists. For some special structures, such as the Maeslandt storm surge barrier in the Netherlands, designers have to consult specialists in the handling of statistical data. He notes that the Eurocodes give a number of tools for dealing with accidental loads but that most failures in practice occur despite the application of the codes. The use of codes should be embedded in a careful strategy for managing risks as recognised in the Eurocode itself. Reliability calculations done for the sake of the record only are almost useless.

The issues mentioned by Ellingwood regarding the acceptability of structural reliability are faced head on by Nethercot in a personal contribution to the debate. He states that enthusiasts of the theory are found almost exclusively in the academic community whereas reaction in the structural engineering community is decidedly mixed. Enthusiasts see reliability theory as a key tool, whereas practitioners see it largely as an irrelevance. He asks if this difference matters and, if it does, can it be bridged? He recognises the theory as a structured way of thinking through the problem by helping to identify key influences and potential contradictions. He identifies part of the tension as arising from confusion between the scientific method and the practice of engineering. He also highlights the importance of low chance/high consequence events and the need for a better theoretical understanding of robustness. He concludes by saying that engineers justify their work by saying they have used best practice and that is underscored by the presence of reliability theory.

Three leading thinkers about the theory of structural risk— Brown, Elms and Melchers—take the debate a stage further. Their contribution is particularly important in the maturing process of the subject. They begin by pointing out that structures seldom fail, so it is reasonable to conclude that they are safe. The process of achieving safety and the process of assessing whether safety has actually been achieved are, however, very different issues. They discuss this through the four related topics of responsibility, failure, uncertainty and decision. They argue that recent codes have focussed only on one aspect of uncertainty, which they call technical uncertainty, with little attempt to address non-technical matters that have been shown to be the origins of most failures. They argue that the decision process is one of 'satisficing' rather than optimising with a process, which is less

rational than is generally supposed and requires a broader outlook from structural engineers.

The final paper is by the guest editor Blockley and is an attempt to begin to show how the need for structural engineers to think more widely can be achieved. A key difficulty is how to manage in a formal way the integration of information and evidence from many disparate sources. The approach suggested is based on systems thinking. The first distinction is between the 'hard physical' systems of traditional engineering science and the 'soft people' systems of engineering management. They can be integrated by focussing on the processes, which represent how physical objects behave and what people do. Measures of evidence called 'Italian flags' to integrate disparate evidence are introduced. Previously they have only been used qualitatively. In the paper a new quantitative theory is introduced.

The emergent themes through all of the papers are

(a) the need to develop a much better theoretical understanding of robustness to deal with the risks of unexpected low chance/high consequence events
(b) the explicit recognition of incompleteness, that is our inability to predict the precise future state of a structure and the possibility of totally unforeseen events
(c) a consequent emphasis on the management of the risks to a structure all through its life cycle; a process in which prediction has a major role rather like the 'observational method' used in geotechnics
(d) the need to recognise the structure as part of a complex, 'softer', wider context which has yet to be considered theoretically in the way that can underpin the issues that practitioners are required to face
(e) a much wider debate and deeper understanding of the relationship between the scientific method and engineering practice which can facilitate the maturing process
(f) the development of techniques to allow for the formal integration of evidence from many disparate sources which will facilitate clear audit trails and transparency of decision making.

In the view of the guest editor (but again not necessarily the guest contributors) the 'systems thinking' approach (Ref. 1 of the paper by Blockley) has much to offer to the much needed maturing of structural risk and reliability theory into mainstream engineering practice. I hope this volume has helped to nudge this process a little further along.

Proceedings of the Institution of
Civil Engineers
Structures & Buildings 161
August 2008 Issue SB4
Pages 171–181
doi: 10.1680/stbu.2008.161.4.171

Paper 700051
Received 13/09/2007
Accepted 13/03/2008

Keywords: buildings, structures &
design/conservation/seismic
engineering

Alberto Bernardini
Associate Professor, Department of
Structural and Transportation
Engineering, University of Padova,
Italy

Sergio Lagomarsino
Professor, Department of Civil,
Environmental and Architectural
Engineering, University of Genoa,
Italy

The seismic vulnerability of architectural heritage

A. Bernardini and S. Lagomarsino

An evaluation of the seismic vulnerability of our architectural heritage is a critical task in the planning of retrofits to buildings in seismic zones. Surviving European monumental buildings are embedded in an urban environment of traditional buildings, the preservation of which is also important. The methodology suggested in this paper is based on the European macroseismic scale in which the classes of vulnerability summarise the observed damage to types of building subjected to past earthquakes. These allow for coherent measuring of the expected behaviour of buildings designed according to modern rules for seismic protection. Criteria are presented for the classifications founded on inventories of buildings based on information varying from poor statistical data about typological parameters of buildings to detailed descriptions of single monuments. Some applications to the safety of villages in the Province of Imperia (Liguria, Italy) are given. Vulnerability curves of monumental structures have been derived from a statistical analysis of damage observed after recent Italian earthquakes. Churches appear to be particularly vulnerable.

NOTATION

D_k damage degree k
I macroseismic local intensity, generally according to the European scale EMS 98
p_k probability of the degree of damage k
Q ductility index
V vulnerability index
α membership level of a fuzzy set (α-cut)
μ mean or expected value

I. INTRODUCTION

Interest in the conservation of our architectural heritage in Europe is very strong. Attention is not just restricted to monumental buildings but also includes ordinary buildings in historic centres. The different sources of risk that threaten a historical structure include: decay of materials; modifications to the structure and/or to the immediate environment; and natural events. Earthquakes, however probably represent the main cause of damage and ultimate collapse. Large areas of Europe have a high level of seismic hazard and ancient masonry structures are vulnerable.

As there are a large number of buildings and monuments in historical centres and urban areas, there is a need for procedures based on simplified models that require few homogeneous data. The methodology described in this paper provides a unified and coherent approach to the vulnerability of a complex system of historical centres and single monuments on a regional scale.

A risk analysis must consider the effects on people (homeless, injured), the losses (economic, but also cultural) and the social impacts. The first step of a vulnerability analysis is the evaluation of the physical damage to non-structural and structural elements.

A macroseismic approach is preferred to a mechanical one. The level of information on any single structure is often rather poor, so identifying a reliable mechanical model is a difficult task. The macroseismic approach, on the other hand, clusters structures into vulnerability classes. These are based on seismic damage expressed as a damage probability matrix, which takes into account observations on buildings stricken by past earthquakes.[1-3] For ordinary buildings, the vulnerability method has been derived directly from the European macroseismic scale (EMS).[4]

In the case of monumental structures, vulnerability curves have been obtained from the statistical analysis of data from the damage assessment after the recent earthquakes in Italy, from Friuli (1976) to Molise (2002). In particular, much of the information is related to churches, which appear to be one of the most vulnerable types of building. The prediction obtained from this methodology is, at best, approximate and can only be used as a first assessment, while more detailed information is required to identify weaknesses in individual historical buildings. For a large population of monuments (for example churches), however, a statistical derivation of expected damage is very important, for example in planning systematic retrofitting interventions at national or regional level.

The layout of the current paper is as follows. First a general macroseismic approach is defined and justified in the European context. The suggested procedures are then more specifically and separately discussed for the general built environment and monumental structures, particularly churches. Finally a case history in the Imperia Province is presented.

2. A MACROSEISMIC APPROACH TO VULNERABILITY

Observations and written records of damage to buildings stricken by past earthquakes are of great utility in reconstructing the historic seismicity of any country, and hence to forecast future seismic hazards.

For example in Italy the catalogue DBMI04 (htpp:// emidius.mi.ingv.it/DBMI04/) contains 58164 observations related to 14161 sites and 1041 different seismic events from 217 BC to 2002 AC.

Since the end of the 19th century observations have been summarised by a synthetic measure: a number in the range from 1 to 9 (Mercalli 1902), 10 (De Rossi 1873) and the 12 levels of Cancani (1904) and Sieberg (1913).[5] For intensities greater than 5 this last scale (and the successive versions of the macroseismic scales) is defined using the observed level of damage to buildings and structures.

Any estimation of the local intensity of an earthquake from recorded damage is, of course, affected by the building types and their vulnerability. Successive versions of the scale therefore generally distinguish damage to different types of buildings. For example the Sieberg scale (presently known as the Mercalli–Cancani–Sieberg scale) distinguishes between 'solidly built mid-European houses', 'poorly built houses', 'earthquake-proof buildings, such as most of the stone and wooden houses in Japan', 'wood and wattled buildings', 'framed buildings', 'churches, minarets of mosques in ...southern Europe', 'church towers and factory chimneys', 'bridges', 'dams', 'pipes in the grounds' and so on.

The degree of damage is qualitatively described, for example, by 'cracks in the plaster', 'cracks in masonry', 'collapses of walls', and the percentage of damaged buildings can be 'none', 'in some cases', 'few', 'a large number', 'most'. Sieberg[6] suggests more precisely in some cases 'nearly a quarter', 'nearly half', 'nearly three-quarters'.

Similar measures of damage to types of buildings can be found in the different versions of the Medvedev–Sponheuer–Karnik (MSK) scale and in the Modified Mercally (MM) scale, extensively used in the last century in Europe and North America respectively.

Despite the qualitative aspect of the recorded information and

the great uncertainty about their reliability, acknowledgement of the historic seismicity helps and confirms the evidence given by the geological survey and related models to estimate the probability of earthquakes of different intensities in the area.

The more recent EMS 98[4] scale introduced further improvements.

(a) The overall damage to buildings, both masonry and reinforced concrete (RC) buildings, has been defined through a scale of six qualitative judgements (from 'no damage' to 'total collapse'), taking into account the severity of damage to structural and non-structural elements.
(b) The vulnerability of a building has been summarised not by the type but by its membership in a well-defined 'class of vulnerability': from A–the most vulnerable–to F–the least vulnerable–to cover the full range of the expected behaviours from traditional non-engineered structures to structures designed for seismic protection. These behaviours are expressed as probabilities of the different degrees of damage conditional on the macroseismic intensities in damage probability matrices (DPM).
(c) Qualitative rules have been suggested to select the most appropriate or probable class for each type and the classes of possible behaviours depending on the quality of the specific building.

The 'vulnerability classes' allow the relative seismic vulnerability of buildings of different types (e.g. a stone masonry building and a RC framed construction) to be compared through the corresponding DPMs.

The definitions suggested by the scale (Table 1) do not give precise implicit values for the DPM. As discussed in section 3, however, it is possible to derive imprecise interval probabilities compatible with the definitions and therefore to estimate bounded intervals of particular functions of the damage (e.g. victims or unusable buildings as a consequence of the earthquake).

In the last 50 years, the widespread distribution of instruments that provide direct records of the strong motions owing to earthquakes apparently seems have reduced the utility or the significance of the macroseismic scale and therefore of an observational approach. For example the Hazus procedure developed by Federal Emergency Management Agency (Fema)[7] suggests a simplified mechanical approach, based on capacity

D_k/I	$k = 0$	$k = 1$ (negligible to slight)	$k = 2$ (moderate)	$k = 3$ (substantial to heavy)	$k = 4$ (very heavy)	$k = 5$ (destruction)
5		Few A or B				
6		Many A or B, Few C	Few A or B			
7			Many B, Few C	Many A, Few B	Few A	
8			Many C, Few D	Many B, Few C	Many A, Few B	Few A
9			Many D, Few E	Many C, Few D	Many B, Few C	Many A, Few B
10			Many E, Few F	Many D, Few E	Many C, Few D	Most A, Many B, Few C
11			Many F	Many E, Few F	Most C, Many D, Few E	Most B, Many C, Few D
12						All A or B, Nearly All C, Most D or E or F

Table 1. Linguistic frequencies of damages for vulnerability classes and macro-seismic intensity according to EMS 98 scale[4] D_k ($k = 0...5$) are the damage degrees defined by the EMS-98

curves and fragility curves, to evaluate the probability of reaching or exceeding specific damage limit states from a given ground motion.

It is, however, apparent that

(a) fragility curves well fitted to the seismic behaviour of modern construction types cannot easily be extended to the Italian (or more generally European) context
(b) a correlation between mechanical and macroseismic approaches based on EMS 98 is possible[8] and can be used for reciprocal calibration of the models
(c) the definitions of the classes of vulnerability (and the implicit related DPMs) given by the macroseismic scale are supported by the available statistical information about the seismic behaviour of traditional non-engineered buildings in Europe through the long sequence of earthquakes of different intensities over the past centuries.

3. VULNERABILITY OF THE BUILT ENVIRONMENT

Historical construction in Europe, particularly in Italy, is characterised by many thousands of small or middle-sized town centres with pre-existing antiquities. The actual configuration was created in the Middle Ages, in the Renaissance or in the Baroque age, sometimes following disastrous natural earthquakes. Many of these historical centres are presently enclosed by modern residential quarters, industrial and commercial areas; the original architecture and urban shape have, however, been generally maintained, although modified by the functional interventions required by modern standards of life. The cultural and economic importance of preserving this heritage for future centuries is largely recognised, although unfortunately not always confirmed in practice.

Traditional masonry buildings are generally grouped in complex interacting groups along networks of narrow streets, whose intricate plans fits the topography of the site and/or the cadastral subdivisions of the properties. During the last century many buildings have been modified by inadequate efforts at restoration, which often use heavy RC slabs – instead of maintaining, and eventually strengthening, the original wooden floors – enlarge doors and windows at the ground storey and increase the number of storeys over the same sub-structure and foundation.

The increased seismic vulnerability of these building types has been demonstrated by the extensive damage and high number of victims recorded after the last two strong earthquakes in Italy (Friuli 1976 and Irpinia 1980). Moreover middle-intensity earthquakes (Umbria-Marche 1997) can also cause victims, together with a disproportionately very high economic damage.

The effective knowledge of such complex systems is often restricted to extremely poor quantitative data (for example overall volume, number of storeys) and qualitative information (for example judgements on the original quality of masonry works and on actual maintenance conditions).[9] In Italy the decennial survey of the National Agency for Statistics (ISTAT) gives information of this nature for the entire stock of residential buildings.

In other cases the survey is much more detailed,[10] describing the characteristics of soils and foundations, connections between walls and floor slabs and between orthogonal walls, and experimental evaluations of the mechanical strength of masonry. On occasions a numerical model is used to estimate fragility curves of a building or of a well-defined building type.[11]

In all cases, however, the available information can be used to estimate the vulnerability class of the building or the homogeneous group of buildings, or at least the probability of membership to some vulnerability classes characteristic for the specific general typology (for example the typology of clay masonry buildings). Taking into account that the macroseismic scale EMS 98 suggests values of such probabilities for some general typologies of buildings, a Bayesian updating[12] could be used when some modifiers of the seismic behaviour could be ascertained for the actual building or group of buildings.

For the effective evaluation of the expected frequencies of the different degrees of damage to buildings, conditional on the macroseismic intensity, knowledge of the DPM implicit in the scale is required. Despite the fact that the qualitative definitions for the frequencies of degrees of damage are incomplete and sometimes are not congruent, the theory of fuzzy sets and random sets can be used to derive a fuzzy version of the DPM.[12,13] Numerical results are described – for every degree of damage, vulnerability class and macroseismic intensity – by 'interval probabilities', depending on a parameter of credibility variable from 0 to 1 based on the α value of the α-cuts of the fuzzy sets given by the scale.

For example in Fig. 1 interval probabilities are displayed for the most vulnerable class A, macroseismic intensities 6 to 9 and the two extreme values (0 and 1) of the α parameter. Moreover a particular probability distribution ('white'[13]) for the damage degrees is displayed. This gives a central value corresponding to the expected percentage of any degree when the probabilistic assignments of the α-cuts are uniformly distributed on the intervals. For example for 'Few' the white value is 7·5%. In the same figure the binomial distributions of the degree of damage for the most vulnerable class of buildings stricken by the Irpinia earthquake (1980; magnitude (M) = 6·7) in the areas with equal estimated MSK intensity can be compared with the implicit EMS 98 DPMs. Although for low intensities some overestimation of the damage seems evident, for middle and high intensities the survey is coherent with the bounds and well approximated to 'white' distributions. A comparison of the EMS 98 and MSK definitions suggest that when the intensity is greater than 5 then the scales can be assumed to be equal.

A useful parametric representation[12] of the upper/lower and white cumulative probability functions of the damage implicit in the EMS 98 can be given through a family (depending on α parameter) of discrete beta distributions with mean value μ_D correlated to the intensity I and a 'vulnerability index' V independent by I through the following relation

$$\mu_D = 2\cdot5 + 3\tanh\left(\frac{I + 6\cdot25\,V - 12\cdot7}{3}\right) \quad 0 \leqslant \mu_D \leqslant 5$$

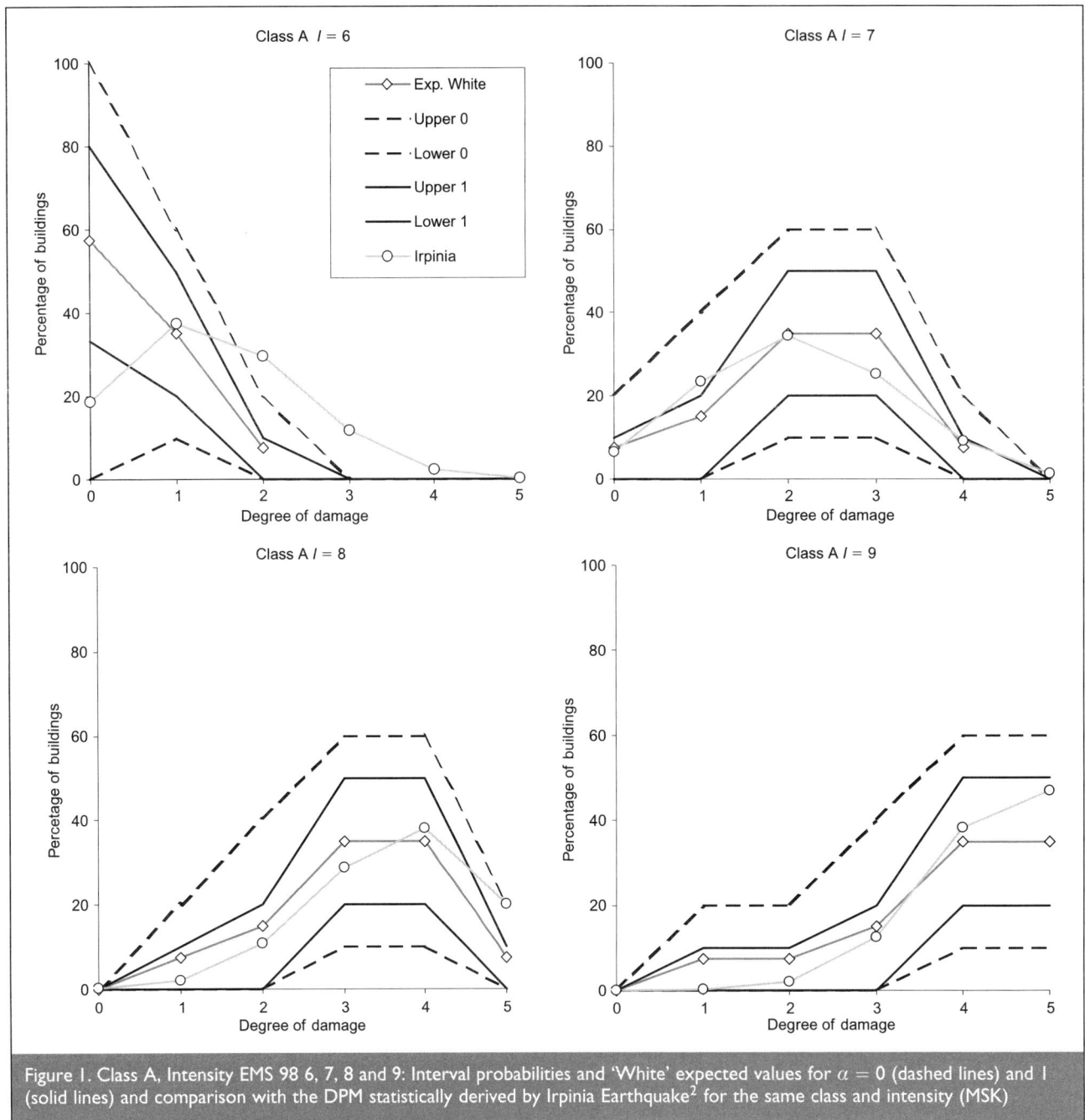

Figure I. Class A, Intensity EMS 98 6, 7, 8 and 9: Interval probabilities and 'White' expected values for $\alpha = 0$ (dashed lines) and I (solid lines) and comparison with the DPM statistically derived by Irpinia Earthquake[2] for the same class and intensity (MSK)

Table 2 gives the white values of V for the 6 EMS 98 classes. When for a specific constructional type the probabilistic assignment of the classes has been evaluated (for example through the Bayesian approach mentioned in section 2), corresponding upper/lower and white values of V can be evaluated.

4. VULNERABILITY OF MONUMENTAL BUILDINGS

The seismic vulnerability of monumental buildings is not explicitly contained in the definition of the modern macroseismic scales as they refer to ordinary multi-storey buildings. The usual approach to the conservation of historical buildings depends on a detailed analysis of each building.

The Italian experience in the last 30 years has shown, however, that the number of monumental buildings, in particular churches and palaces, is rather high and so there is the need for statistical methods to evaluate seismic risk at a regional scale.

A vulnerability model can be derived through an observational approach using a statistical analyses of data obtained during the damage assessment.

Class	A	B	C	D	E	F
V_0	0·88	0·72	0·56	0·40	0·24	0·08

Table 2. White values of the vulnerability index V for the EMS 98 vulnerability classes

5. DAMAGE ASSESSMENT

An earthquake is a real test of the actual vulnerability of structures. Post-earthquake damage assessments are a remarkable source of information. This activity is coordinated in Italy by the Civil Protection Department. Official survey forms are used to obtain systematic and objective data to

(a) verify if the structure is fit to use in the emergency (green, yellow or red tag)
(b) decide for provisional interventions (propping, hooping), in order to prevent from further damage owing to after shocks
(c) evaluate economic losses, to be charged to insurance companies or public funding
(d) make a preliminary diagnosis of the building performance, to design the proper repairing and strengthening interventions.

In the case of monuments, the first problem is how to define the damage. For many types of historical buildings, such as palaces, villas or convents, the EMS 98 definition in terms of six damage degrees can be used.

Starting from the Friuli earthquake (1976), damage to churches has been systematically assessed through an approach that considers a subdivision of the structure into sub-structures referred to as 'macro-elements'. These macro-elements usually correspond to well-defined parts, which behave almost independently in the case of an earthquake. Typical examples of macro-elements are the façade, the nave, the triumphal arch, the dome, the apse and the bell tower. Observations show recurrent behaviours, according to damage and collapse mechanisms.

In particular, the damage to more than 2000 churches after the Umbria and Marche earthquake in 1997 has been assessed.[14] Eighteen indicators are used, each one representative of a possible collapse mechanism in a macro-element (Fig. 2). During the survey, a degree of damage is assigned to each one of the possible collapse mechanisms and finally a damage index is defined for the whole church.

A statistical analysis of the data results in a DPM for the churches. This contains the probability p_k of five damage levels D_k for four different values of the macroseismic intensity. The damage levels are defined according to the modern macroseismic scales, in particular the EMS 98 (see Table 1): from $k = 0$ (no damage) up to $k = 5$ (destruction). For each level of intensity, the mean damage grade μ_D is given by

$$4 \quad \mu_D = \sum_{k=1}^{5} k p_k$$

The damage histograms displayed in Fig. 3 are well fitted by the binomial distribution, which is defined only by one parameter, the mean damage grade μ_D (or the binomial coefficient $\mu_D/5$)

$$3 \quad p_k = \frac{5!}{k!\,(5-k)!} \left(\frac{\mu_D}{5}\right)^k \left(1 - \frac{\mu_D}{5}\right)^{5-k}$$

Figure 3 shows the gradual and regular increase of the mean damage μ_D with the intensity. In recent years, damage to churches surveyed after earthquakes in other regions confirm the relationships obtained from the Umbria and Marche data.

The DPM, which considers all the churches as a homogeneous type, characterises the average vulnerabilities. The survey form, however, allows the specific vulnerability of each church to be taken into account and evaluated through a vulnerability index. By dividing the sample of the churches that suffered the same seismic intensity into smaller samples characterised by ranges of the vulnerability index V, a more refined correlation may be set up,[15] which also considers this intrinsic parameter of the church

$$4 \quad \mu_D = 2.5 + 3 \tanh\left(\frac{I + 6.25\,V - 12.7}{3.7}\right) \quad 0 \leqslant \mu_D \leqslant 5$$

where V assumes values between 0·65 and 1·2.

The vulnerability curve (4) is a model through which the probability damage distribution of a church can be evaluated and in which the vulnerability index V has been obtained by a proper survey, as a function of the macroseismic intensity I. The model is very similar and coherent with the model presented above (section 3) for current building typologies and derived from the EMS 98 qualitative definitions.

Among the different types considered in EMS 98, massive stone masonry types may be, on average, associated with monumental palaces, because their construction is typically characterised by good-quality materials and craftsmanship. For massive stone buildings the white value V_0 has been estimated equal to 0·62, and the vulnerability index usually varies, according to the vulnerability scores (the modifiers), between 0·49 and 0·79.

It is worth noting that the vulnerability curves of churches (4) and palaces (1) are similar, with the exception of the denominator, which controls the rate of increase of the damage with the intensity. This is defined as the ductility index, Q. The vulnerability curves of the churches and of the monumental palaces are compared in Fig. 4. Owing to higher values, on average, of the vulnerability index, churches turn out to be more vulnerable for the lower intensities; actually, in the case of minor earthquakes in Italy, the churches always exhibited a higher level of damage. The higher ductility of churches ($Q = 3.7$ for churches, $Q = 3$ for palaces) indicates that damage is similar for higher intensities .

6. VULNERABILITY MODEL

A vulnerability model for historical buildings, suitable for application at a regional scale, has to refer to a class type. There is such a wide variety of artefacts within our cultural heritage, whether owing to geography, architectural style or age of construction, that this classification is not a straightforward task. For the sake of a simplified seismic evaluation, however, a class type is made by gathering into groups the structures that are similar with reference to the use, architecture and possible seismic behaviour.

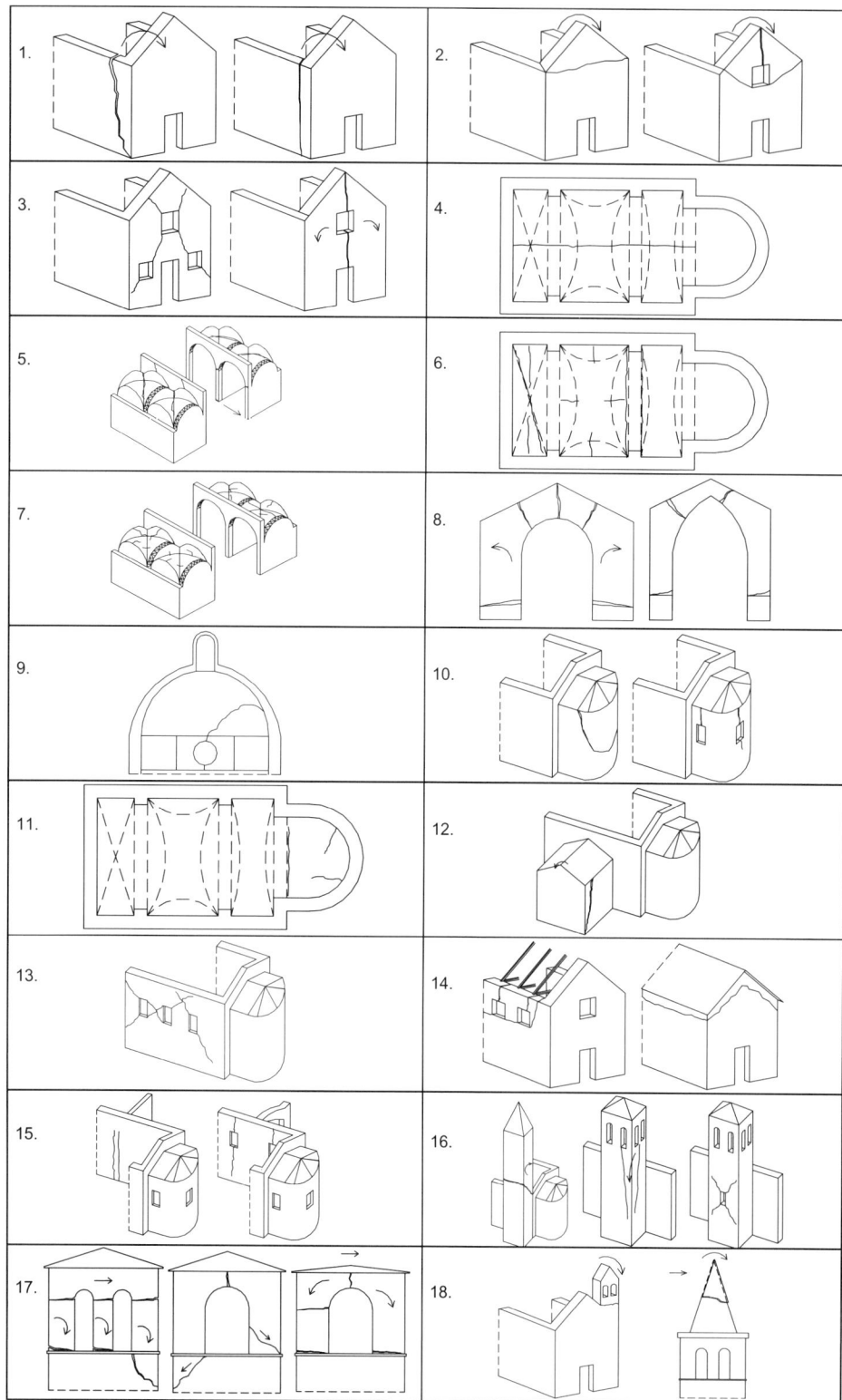

Figure 2. Damage mechanisms in the church macro-elements

A suggested classification for monuments in European countries, is the following: palace, church, monastery/convent, mosque, tower, obelisk, theatre, castle, triumphal arch and arch bridge.

The vulnerability model gives the mean damage grade as a function of the macroseismic intensity. It is defined by two parameters: the vulnerability index V and the ductility index Q (see equations (1) and (4)). The white vulnerability index V_0 for monumental palaces and churches can be deduced respectively by the implicit definition in EMS 98 and a statistical analysis of observed damage data, as specified in section 5. For towers and castles, vulnerability parameters have been obtained[16] by damage data from other Italian earthquakes (Irpinia 1980 and Salò 2004). Reference values for the other monumental types may be deduced from them, according to expert judgement and taking advantage of some observed data on each typology.[17] The values in Table 3 can be used as a reference.

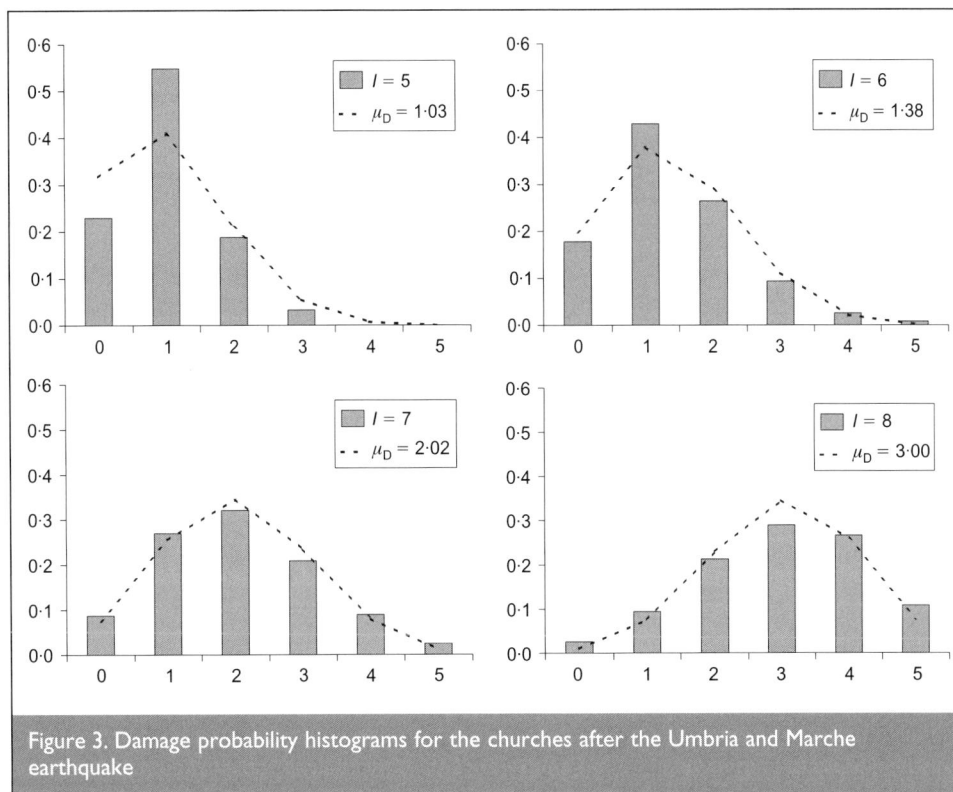

Figure 3. Damage probability histograms for the churches after the Umbria and Marche earthquake

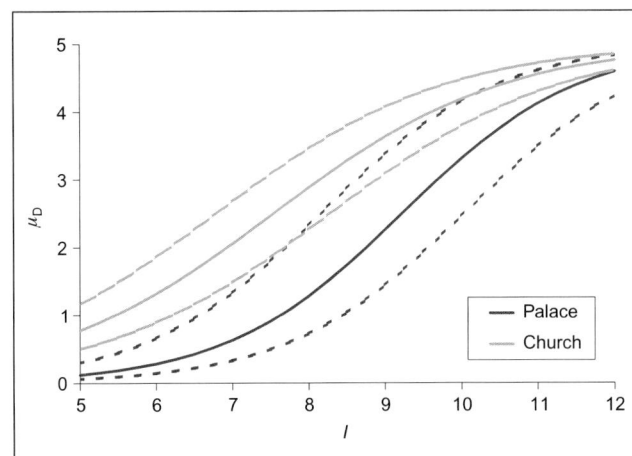

Figure 4. Vulnerability curves of palaces and churches (White value in solid line; upper and lower bounds in dashed lines)

Typology	V_0	Q
Palace	0·62	3·0
Church	0·89	3·7
Monastery/Convent	0·74	3·0
Mosque	0·81	3·4
Tower	0·78	2·6
Obelisk	0·74	3·7
Theatre	0·70	3·0
Castle	0·54	2·6
Triumphal arch	0·58	3·4
Arch bridge	0·46	3·0

Table 3. Parameters for the macroseismic vulnerability model of historical constructions

7. DAMAGE SCENARIOS

Once the hazard scenario is known, a damage risk for each monument can be readily evaluated (damage scenarios). The mean damage grade μ_D, given by equation (4) with the proper value of the ductility index Q and $V = V_0$, represents a synthetic meaningful parameter for the damage scenario; Fig. 5 shows the mean vulnerability curves for the different monumental typologies.

Clearly a vulnerability index assigned to a monument by a typological classification represents an average value, which does not take into account the distinctiveness of the single building and does not allow the most vulnerable structures among buildings of the same type to be singled out. To refine the vulnerability assessment, a quick survey is at least necessary, collecting further relevant parameters such as: state of maintenance, quality of materials, structural regularity (in plan and in elevation), size and slenderness of relevant structural elements, interaction with adjacent structures, presence of retrofitting interventions, site morphology.

Vulnerability scores V_j may be awarded to each one of the above mentioned parameters and the vulnerability index of each monument may be refined, modifying the typological value:

$$5 \qquad V = V_0 + \sum V_j$$

where the summation is extended to all the available modifiers. The vulnerability scores may assume different values for different types, and some information may be relevant only for some types. The choice of the specific vulnerability parameters has been made empirically, on the basis of the observed damage, according to similar procedures proposed in different countries for buildings.[18,19] The values of the modifying scores has been tuned, for churches and palaces, by the statistical analysis of data that come out from damage assessment. Some reference values are proposed in Table 4.

8. A CASE HISTORY: THE IMPERIA PROVINCE (ITALY)

The Province of Imperia is located around the Mediterranean Sea, on the border between Italy and France. About 210 000 residents live in an area of 1154 km²; 56 000 in the main town, Sanremo. A research project, funded by the Italian Civil Protection Department, identified 1365 monumental buildings and 96 historical centres;[20] described in a geographical

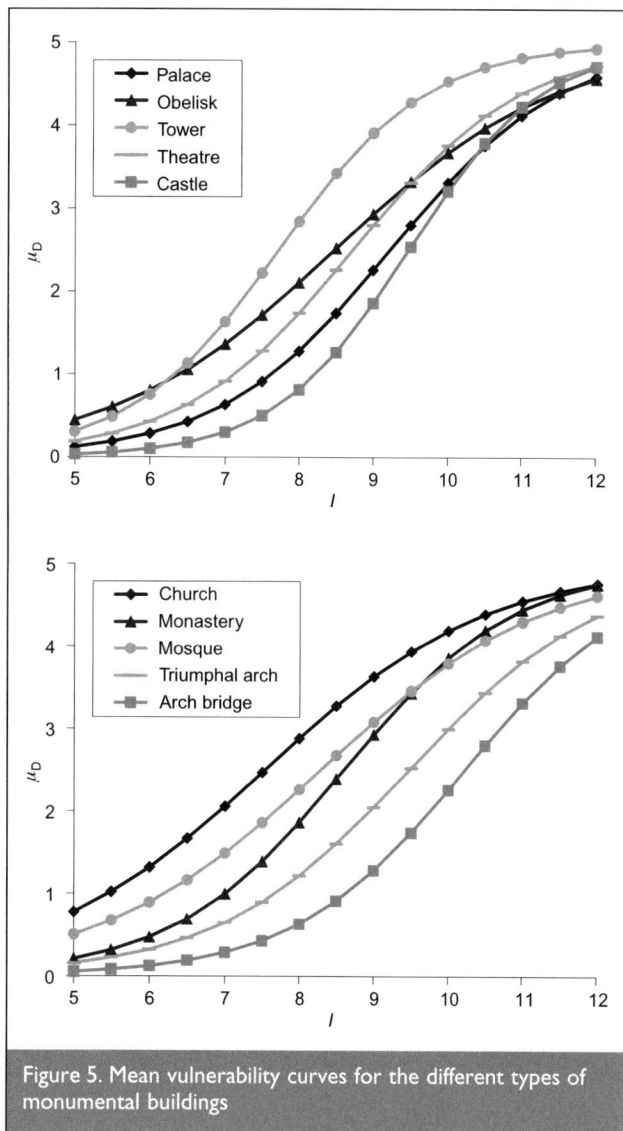

Figure 5. Mean vulnerability curves for the different types of monumental buildings

number of inhabitants. Some historical centres are very small and scarcely populated. Table 6 shows the number of monuments by type, distinguishing between those included in historical centres, other urban centres and isolated areas.

In an analysis of the seismic risk analysis to our cultural heritage, the consequences of a scenario event need to be evaluated. This analysis must consider not only the physical damage but also the losses in terms of historical, architectural and artistic value. To this end a model has been proposed which defines the significance of these assets, through a score assigned to each historical centre. The score, ranging from 0 to 10, is obtained by adding two contributions

(a) the presence of monuments inside the centre
(b) the urban and landscape characteristics.

These contributions are defined through both quantitative and qualitative parameters.

For monuments, a score up to 2 is assigned by a function that considers the summation of the ratios between the number of monuments in the historical centre and the total number in the considered area, separately for the different types of monuments (Table 3). Regarding the qualitative aspects (essentially architectural and artistic assets), a score of up to 3 is assigned. This is based on the number of citations in a tourist guide,[21] drawn up by authoritative fine-arts historians and used by the Ministry of Culture for the Italian catalogue of monuments, aimed at the conservation against natural and anthropic risks (called 'Carta del rischio').

For urban characteristics, a score of up to 2 is assigned by a function that considers the ratio between the number of ancient buildings and the total number buildings in the historical centre. Regarding the qualitative aspect, a score of up to 3 is assigned based on three factors: historical importance, the urban environment and the quality of harmony with the landscape, all obtained using the above-mentioned tourist guide.

information system (GIS), available on line (http://adic.diseg.unige.it/gndt-liguria/index.html). Table 5 shows some data of historical centres, in particular the number of buildings, the percentage of old masonry buildings and the

Parameter	V_j
State of maintenance	Very bad (0·08); bad (0·04); medium (0); good (−0·04)
Quality of materials	Bad (0·04); medium (0); good (−0·04)
Planimetric regularity	Irregular (0·04); regular (0); symmetrical (−0·04)
Regularity in elevation	Irregular (0·02); regular (−0·02)
Interactions (aggregate)	Corner position (0·04); isolated (0); included (−0·04)
Retrofitting interventions	Effective interventions (−0·08)
Site morphology	Ridge (0·08); slope (0·04); flat (0)

Table 4. Reference values for vulnerability scores V_j of the main parameters

	Total	Min	Max	Mean
Number of buildings	16 977	18	682	177
Percentage of masonry buildings	81·2%	17·7%	100%	—
Number of inhabitants	47 340	15	5250	493

Table 5. Characteristics of the 96 historical centres in the Province of Imperia

Typology	Total number	Within the 96 historical centres	Within other urban centres	Outside urban centres
Palace	396	198	128	70
Church	713	361	80	272
Monastery/convent	21	10	9	2
Tower	77	27	23	27
Theatre	8	4	4	0
Castle	72	40	4	28
Arch bridge	78	25	2	51
Total	1365	665	250	450

Table 6. Numbers of monuments in the Province of Imperia

The hazard scenario has been derived from the main historical earthquake in the Western part of the Ligurian Region that occurred on 23 February 1887[22] and which produced very serious damage and many victims along the coast of Piedmont and France. This event is a reference point in the seismic history of the region, as some effects are still present. Some monuments were not restored and survive as ruins, while an historical centre, Bussana, was abandoned and only now is to be rehabilitated. The maximum intensity was degree 10 and damage was considerable with more than 600 victims within a population, at that time, of 49 000. An intensity scenario has been produced, for the identified epicentre and by means of a proper attenuation law, obtaining intensities in the area between 7·5 and 9 (degree 10, estimated in 1887 only in two small centres, has not be considered).

The seismic vulnerability has been evaluated, both for the monumental buildings and for the ordinary buildings in the historical centres. A damage scenario for the cultural heritage in the Imperia Province has also been produced.

As regards current buildings, Fig. 6(a) shows the histogram of the mean damage grade in the 96 historical centres. A mean damage grade of $\mu_D = 2$ is foreseen on average in the whole area. Fig. 6(b) shows the histogram of damage grade for the masonry buildings in the Imperia Province. The mean expected number of collapsed buildings is 217 (1·3% of the total number of buildings), while there are 3154 unusable buildings.

Table 7 shows the distribution of the foreseen mean damage grade for different monument types. It is worth noting that churches turn out to be not only the most numerous but also the most vulnerable. For 25 churches a mean damage level of 4 is expected, which implies a high probability of collapse (local or global). This result agrees with the observed effects produced by the 1887 earthquake. Towers also appear to be very vulnerable, while damage to masonry arch bridges is negligible.

The physical damage to buildings and monuments has been related to the loss of cultural significance (Fig. 7(a)). The cultural loss for the monumental buildings is more sensitive to physical damage. This is because even low damage levels may ruin frescoes or other high-value artistic assets. The loss rate function is assumed, considering a binomial distribution, as the percentage of damage grade 5 for ordinary buildings, and the percentage of damage greater than 4 for monuments.

Figure 7(b) shows the histogram of cultural significance

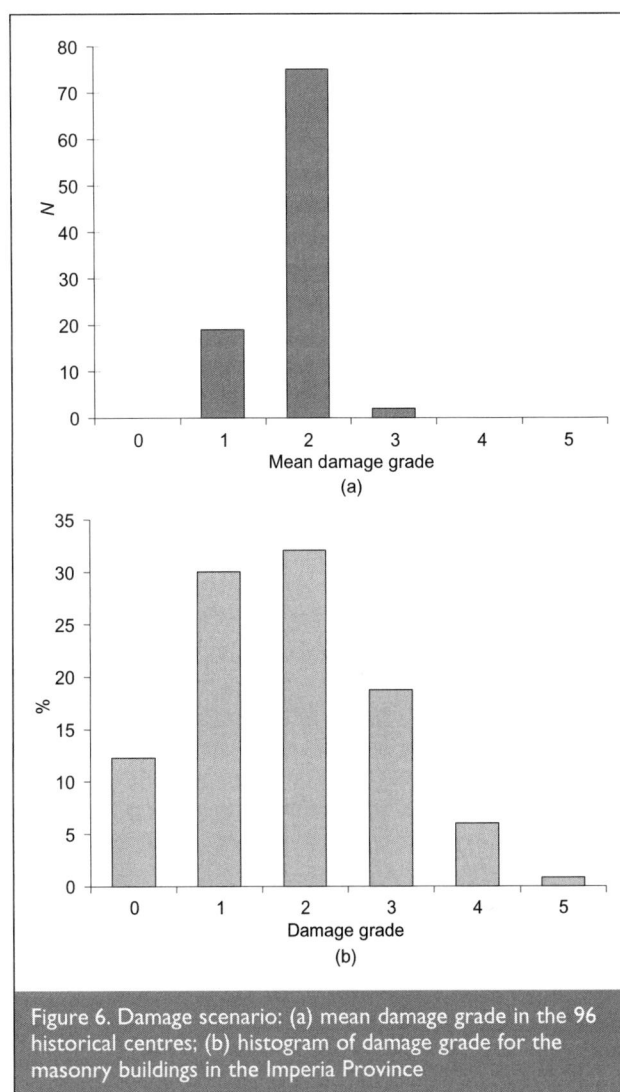

Figure 6. Damage scenario: (a) mean damage grade in the 96 historical centres; (b) histogram of damage grade for the masonry buildings in the Imperia Province

(defined above) of the historical centres in the Imperia Province, before and after the earthquake.

9. CONCLUSIONS

A knowledge-based methodology has been demonstrated as particularly useful to forecasts of the seismic behaviour of the historic constructional heritage of the Italian towns and villages. The original knowledge for current buildings is given by the definitions and qualitative frequencies of damage suggested by the macroseismic scale EMS 98 for six vulnerability classes. Different structural types are characterised by probabilities of membership within the classes.

Typology	No damage	Slight damage	Moderate damage	Heavy damage	Very heavy damage	Collapse
Palace	0	106	233	55	2	0
Church	0	0	235	453	25	0
Monastery/Convent	0	0	13	7	1	0
Tower	0	7	32	33	5	0
Theatre	0	4	4	0	0	0
Castle	9	40	14	9	0	0
Arch bridge	30	47	1	0	0	0

Table 7. Distribution of mean damage values in the different typologies of monuments

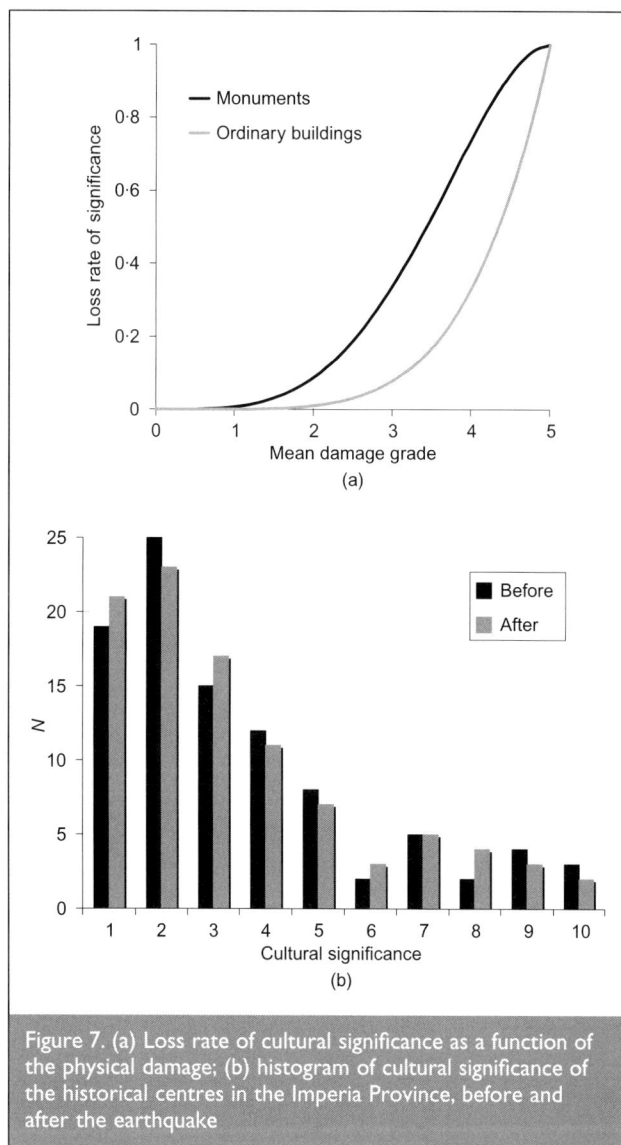

Figure 7. (a) Loss rate of cultural significance as a function of the physical damage; (b) histogram of cultural significance of the historical centres in the Imperia Province, before and after the earthquake

high. Nevertheless it is useful in emergency management and in planning retrofitting priorities in risk reduction programs. Moreover Bayesian procedures or expert rules can be used to reduce uncertainty when more detailed information about more homogeneous groups of buildings or also single buildings and monuments becomes available.

A simple numerical parameter, the index of vulnerability, has been used in both cases to obtain a coherent and homogeneous measure of the seismic vulnerability. This parameter appears to be nearly independent by the macroseismic intensity, at least for the typical maximum intensities in Italy and south Europe. Expert methodologies to measure vulnerability, generally suggested for quick survey of buildings in the after-shock time, can therefore be calibrated in the ambit of a probabilistic model describing the relation between damages and macroseismic intensities.

The application to the Imperia Province (Italy) showed that the methodology makes it possible to evaluate scenarios describing both physical damage and loss of cultural significance.

REFERENCES

1. WHITMAN R. V., REED J. W. and HONG S. T. Earthquake damage probability matrices. *Proceedings of 5th European Conference on Earthquake Engineering, Rome*, 1973, pp. 2531.
2. BRAGA F., DOLCE M. and LIBERATORE D. A statistical study on damaged buildings and an ensuing review of the M.S.K76 scale. *Proceedings of 7th European Conference on Earthquake Engineering, Athens*, 1982, 431–450.
3. COBURN A. and SPENCE R. *Earthquake Protection*. Wiley, Chichester, 1992.
4. GRUNTHAL G. European Macroseismic Scale 1998. *Cahiers du Centre Européen de Géodynamique et de Séismologie*, 1998, 15, 1–97.
5. DAVISON C. On scales of seismic intensity and on the Construction and use of Isoseismal Lines. *Bulletin of the Seismological Society of America*, 1921, 11, No. 2, 95–130.
6. SIEBERG A. Geologie der Erdbeben. In *Handbuch der Geophysics* (GUTENBERG B. (ed.)). Gebruder Borntraeger, Berlin, 1930, pp. 550–555.
7. HAZUS *Earthquake Loss Estimation Methodology:Technical and User Manuals*. Federal Emergency Management Agency, Washington, D.C., 1999.
8. LAGOMARSINO S. and GIOVINAZZI S. Macroseismic and mechanical models for the vulnerability and damage assessment of current buildings. *Bulletin of Earthquake Engineering*, 2006, 4, No. 4, 415–443.

The more specific knowledge for particular monumental types has been updated through the observations of the damages described in the available literature and systematically recorded after the more recent Italian earthquakes.

Non-classical methodologies (fuzzy sets, random sets and imprecise probabilities) make it possible to calculate the numerical bounds and central 'white' values of the expected values of damage grades and of any dependent function of interest. When using only information based on typological parameters, the uncertainty about the expected values is very

9. D'AYALA D. F., SPENCE R. S., OLIVEIRA C. S. and POMONIS A. Earthquake loss estimation for Europe's historic town centres. *Earthquake Spectra*, 1997, 13, No. 4, 773–794.

10. DOLCE M., KAPPOS A., MASI A., PENELIS G. and VONA M. Vulnerability assessment and earthquake scenarios of the building stock of Potenza (Southern Italy) using Italian and Greek methodologies. *Engineering Structures*, 2006, 28, No. 3, 357–371.

11. D'AYALA D. Force and displacement based vulnerability assessment for traditional buildings. *Bulletin of Earthquake Engineering*. 2005, 3, No. 3, 235–266.

12. BERNARDINI A., GIOVINAZZI S., LAGOMARSINO S. and PARODI S. The vulnerability assessment of current buildings by a macroseismic approach derived from the EMS-98 scale. *3° Congreso Nacional de Ingenieria Sismica, Girona, Spain,* 2007, pp. 704–718.

13. BERNARDINI A. Random sets of the expected seismic damage to building types. *Proceedings of the 9th ICOSSAR, Roma,* 2005. CD-rom.

14. LAGOMARSINO S. and PODESTÀ S. Seismic vulnerability of ancient churches. Part 1: damage assessment and emergency planning. *Earthquake Spectra,* 2004, 20, No. 2, 377–394.

15. LAGOMARSINO S. and PODESTÀ S. Seismic vulnerability of ancient churches. Part 2: statistical analysis of surveyed data and methods for risk analysis. *Earthquake Spectra,* 2004, 20, No. 2, 395–412.

16. CURTI E., LAGOMARSINO S. and PODESTÀ S. Dynamic models for the seismic analysis of ancient bell towers. *Fifth International Conference on Structural Analysis of historical Constructions* (LOURENÇO P. B., ROCA P., MODENA C. and AGRAWAL S. (eds)). Macmillan, India, 2006, pp. 1067–1074.

17. LAGOMARSINO S. On the vulnerability assessment of monumental buildings. *Bulletin of Earthquake Engineering,* 2006, 4, No. 4, 445–463.

18. BENEDETTI D. and PETRINI V. Sulla vulnerabilità sismica degli edifici in muratura: un metodo di valutazione. *L'Industria delle Costruzioni,* 1984, 18, No. 149, 66–78 (in Italian).

19. FEDERAL EMERGENCY MANAGEMENT AGENCY. ATC-21 *Rapid Visual Screening of Buildings for Potential Seismic Hazards: A Handbook.* FEMA, Washington, DC, 1988. FEMA Report 154.

20. BALBI A., LAGOMARSINO S. and PARODI S. Seismic vulnerability and significance of historical centres and monumental buildings: a case study in the Province of Imperia. In *La Messa in Sicurezza Del Territorio da Eventi Calamitosi di Tipo Naturale* (MINCIARDI R. and UGOLINI P. (eds)). Alinea Editrice, Firenze, Milan, Italy, 2006, pp. 123–144 (in Italian).

21. TOURING CLUB ITALIANO. *Red Guides of Italy: Liguria.* TCI, 1982 (in Italian).

22. TARAMELLI T. and MERCALLI G. Il terremoto ligure del 23 febbraio 1887. *Annali Ufficio Centrale di Meteorologia e di Geodinamica,* 1898, VIII, 331–626 (in Italian).

What do you think?

To comment on this paper, please email up to 500 words to the editor at journals@ice.org.uk

Proceedings journals rely entirely on contributions sent in by civil engineers and related professionals, academics and students. Papers should be 2000–5000 words long, with adequate illustrations and references. Please visit www.thomastelford.com/journals for author guidelines and further details.

Proceedings of the Institution of
Civil Engineers
Structures & Buildings 161
August 2008 Issue SB4
Pages 183–188
doi: 10.1680/stbu.2008.161.4.183

Paper 0700052
Received 14/09/2007
Accepted 28/03/2008

Keywords: buildings, structures &
design/risk & probability analysis/
safety & hazards

Jitendra Agarwal
Senior lecturer, University of
Bristol, UK

Juan England
Senior catastrophe risk
analyst, Willis Research
Network, London, UK

Recent developments in robustness and relation with risk

J. Agarwal PhD and J. England PhD

Structural engineering is a mature discipline, but in practice the nature of demands or actions on a structure remains uncertain. It is difficult to foresee where damage, for example resulting from an accident or material degradation, might occur. A structure that is robust will be able to cope with such unexpected demands without disproportionate consequences. There is, however, no satisfactory measure of robustness: not even a widely agreed definition. In this paper an overview of the current guidelines and methodologies to achieve a robust structure is provided. Most of the analytical studies assume a model of loading, but to achieve robustness there is a need to identify potential weaknesses in the form of a structure. In this respect, structural vulnerability theory can be an effective tool to reduce the risk of propagation of failure owing to unforeseen damage or actions. Similar models are already used in the insurance industry to determine the risk to natural and man-made hazards.

1. INTRODUCTION

Structural engineering as a discipline is well advanced, but the collapse of World Trade Center towers as a result of terrorists' attacks has proved to be a turning point in the way structures are designed and assessed for safety. Since that high-profile event, there have been several studies and a large number of articles have appeared in the literature. Codes of practice have been updated and new guidelines have been formulated. It is the second time that such a shift has taken place: the first was seen after the failure of Ronan Point flats in London owing to a gas explosion. The actions causing the initial damage in these two cases were quite different in their nature and scale but the way initial damage led to the failure of each structure is directly related to the structural configuration, that is how the members were connected. The form of a structure is fundamental to the many theories and methods of structural safety. Some of the methods to ensure safety such as reliability analysis have been around for many years, but methods to examine the robustness are in the early stages of development.

The objective in this paper is to provide an overview of the recent developments for ensuring structural robustness and to examine how these fit within a broader framework of structural risk. Commonly used terms are interpreted differently by many researchers and practitioners, so definitions are provided. In section 3 progressive failure is discussed, including analytical

studies and the factors affecting robustness. A review of current guidelines on robustness is given in section 4. Some of the more recent approaches for robustness, energy methods and structural vulnerability analysis, are examined in sections 5 and 6 respectively. Finally, the important issue of managing structural risks is discussed and the similarities with loss estimation methodologies used in the insurance industry are highlighted.

2. DEFINITIONS

Many terms are interpreted differently by different users; for example, there is often no clear distinction between hazard and vulnerability. The following definitions (see also Blockley[1]) are given as a basis for subsequent discussion and shared understanding.

(a) *Robustness* is an attribute of a system that relates to its ability to fulfil its function in the face of uncertain and adverse conditions such as the loss of a component or abnormal changes in the demands. In the context of structures and this paper, it implies the ability of a structure to avoid disproportionate consequences in relation to the initial damage.

(b) *Structural integrity* is the quality of being whole and sound. At the component level, it means that the material is free from defects. At the system level, it can be thought of as the safe performance of the structure.

(c) *Vulnerability* is the susceptibility of a structure to some action. Generally this susceptibility arises because of an inherent weakness in the structure which could be exploited by one or more potential actions. A structure which is vulnerable in any one way cannot be robust.

(d) *Fragility* implies easily damaged or broken but has developed the specialised technical meaning of the probability of a stated level of damage for a specific hazard.

(e) *Disproportionate collapse* results from small damage or a minor action leading to the collapse of a relatively large part of the structure. It is a measure of the degree of consequential damage.

(f) *Progressive collapse* is the spread of damage through a chain reaction, for example through neighbouring members or storey by storey. Often progressive collapse is disproportionate but the converse may not be true.

(g) A *hazard* is the state of a structure or an action on it with a potential for harm. A corroded member is an example of a hazard which is internal to the structure and an earthquake is an external hazard.

(h) *Reliability* is the probability that a structure or a component will perform satisfactorily over a period of time.

(i) *Risk* is a measure of the chance of something going wrong and the consequences flowing from it set in a context. Risk is always in the future which must be managed. It is not possible to design a structure with 'zero risk'.

3. PROGRESSIVE FAILURE AND ROBUSTNESS

Several studies (e.g. Refs 2–6) have been published on the progressive failure of trusses and frames, but all types of structures are susceptible to progressive collapse. There are three key processes

(a) choice of a member failure model
(b) notional removal of failed member from the analytical model
(c) an analysis of resulting consequences.

The sudden removal of a member causes a change in the geometry of the structure with a consequent release of potential energy. When the change in potential energy exceeds the strain energy that can be absorbed by the damaged structure, then progressive collapse is initiated. This clearly is a dynamic process with internal dynamic loading of the remaining structure which can be modelled using energy principles. The input loading that causes the initial damage is, however, often not included to keep the analysis hazard independent. Few studies include non-linear dynamic analysis and results from a static analysis should be treated carefully as the structure may remain unsafe.

Gross and McGuire[2] studied frames by selectively removing one or more members from a non-linear static structural model. The equilibrium was restored by applying a set of unbalanced forces, given by the forces in the removed members, to the remaining structure. Malla and Nalluri[3] used a member failure model based on the non-linear post-buckling behaviour of a strut and the dynamic effects were modelled using the pseudo-force method. A modified pushover analysis using a vertical load instead of a horizontal load has been used by Warn et al.[7] The response of the structure to the incremental loads helps to trace the stiffness degradation until progressive collapse occurs.

Such studies are aimed at analysing the propagation of damage in a structure under specific loading conditions, rather than identifying where the structure is vulnerable or how robust it is. Of course several such analyses can help to identify a robust structure. A structure which is robust is unlikely to lead to a progressive failure. Robustness can often be recognised, but it is difficult to measure or analyse. It is the ability of a structure not to be damaged disproportionately in the event of an accident, misuse or deterioration. Disproportionate collapse can be avoided by various strategies such as

(a) using a sound and holistic structural concept
(b) paying attention to elements which result in the instability of the whole structure

(c) providing effective measures to stop local damage spreading.

There is broad agreement, as reflected in an Institution of Structural Engineers report,[8] that redundancy, strength, ductility and capacity to absorb energy are key elements to achieve a robust structure.

The issue of robustness should be considered during the early stages of design. An important element is the consideration of how easily the structure could be collapsed with a minimum of effort, were it applied most effectively. This is a mammoth task, given the various possibilities of loads and how these are resisted; however, it relates to structural form, which should incorporate structural redundancy offering alternate load paths. There should be no weak links in the redundant path. Where form is not redundant all elements must be made robust and appropriate safeguards introduced.

There have been attempts to quantify robustness by considering the ratio of some performance measure of a system in its damaged and undamaged state. Meas et al.,[9] for example, have considered (a) system strength and (b) system failure probability. Such quantities capture specific attributes of the system performance for a loading combination. The Standing Committee on Structural Safety (Scoss)[10] recommended the following actions

(a) avoid, eliminate or reduce the hazards to which the structure may be exposed
(b) select a structural form which has low sensitivity to the hazards considered with conscious emphasis on the influence of the structural integrity
(c) provide adequate ductility of the structure for energy absorption
(d) make due allowances during design for the effects of construction tolerances.

These actions recognise that physical means of achieving robustness are important but robustness is considered in a wider sense, which includes the system environment. Some of these issues relate to reducing risk rather than to structural robustness.

4. GUIDELINES ON ROBUSTNESS

Many building regulatory agencies provide guidelines to avoid the potential of progressive collapse and thus improve the robustness of structures. There is, however, no attempt to quantify the robustness of a structure.

In the UK building regulations[11] and British standards,[12] the purpose is to design robust buildings that will sustain localised failure from any action and that will not result in disproportionate consequences. Buildings are classified into four categories (class 1, 2A, 2B and 3) based on the type and occupancy. For each class of building there are specific design criteria to achieve robustness. For class 1 buildings no specific measures other than the usual requirement of horizontal tying are required. For class 2A buildings, also, it is sufficient to ensure the provision of horizontal ties which should be able to resist a tensile force of 75 kN. For class 2B buildings additional provision of vertical ties and if necessary, the design of key elements is recommended. The mechanism of effective tying aims to provide strength and continuity to the structure. In the case of accidental removal of a column, tying can help to achieve a catenary action. On the other hand, key elements are designed to ensure that such members will remain functional during and after an accidental damage. Key elements are identified by checking the stability of the structure on the notional removal of a column. When the removal of a member results in the collapse of more than 15% or 70 m^2 of the floor area of that storey, and at the same time extends further than the immediate adjacent floors, the member is designed as a key element. Class 3 buildings (such as grandstands for over 5000 spectators, buildings over 15 storeys and structures containing hazardous materials) require a systematic risk assessment that considers all possible hazards to the structure. The objective is to determine if there are unacceptable risks and to find measures to mitigate them. The design of the structure is then performed on the basis of the outcome of the risk assessment, including all foreseen hazards, and by satisfying the robustness design criteria for class 2B buildings.

Eurocodes follow a similar approach to the one discussed above. The requirements are set out in EN1990[13] and more detailed guidelines for the design of structures to accidental actions are contained in EN1991.[14] The US approach for mitigating progressive collapse is also based on providing structural continuity, redundancy, and ductility. ASCE-7[15] gives three ways to approach structural design against progressive collapse: indirect method, alternate load path method and specific load resistance method. The so-called indirect method is an implicit design approach which incorporates measures to increase the overall robustness of the structure, for example, through the selection of structural system, structural layout, member proportioning and detailing of connections. The alternate load path method provides a formal check on the capability of the structure to resist collapse following the removal of specific elements. The specific load resistance method is used explicitly to design critical vertical load-bearing components to resist the loads specific to a threat.

In US guidelines,[16] there appears to be some flexibility in the analysis procedure, which could be linear or non-linear, static or dynamic, two-dimensional (2D) or three-dimensional (3D). In a linear analysis, demand capacity ratio (DCR) indicates the magnitude and distribution of potential inelastic demands on the structural elements. US General Services Administration guidelines[16] specify allowable DCR values for different demands, for example flexure, shear. These range from 1 to 3 depending upon the construction type and configuration. If DCR values in flexure exceed the limit, the ends are released thus redistributing the moments. The acceptance criteria for non-linear analyses are based on the ductility and rotation limits rather than the DCR values. Unified facilities criteria[17] issued by the US Department of Defence place buildings into four categories according to the required level of protection. The requirements are similar to those in the UK but a peer review is also required for medium and high levels of protection.

Broadly speaking, there is not a big difference in the approaches adopted by different bodies/agencies to avoid disproportionate collapse. While US General Services

Administration (GSA) procedure follows a direct design approach, United Facilities Criteria (UFC) incorporate both direct and indirect design approaches. The latter are very similar to British requirements for effective tying. Given the evolving nature of terrorist threat, it is impossible to predict what threats may be of concern during the lifetime of a structure. The notional removal of a member and the alternate load path method may identify disproportionate collapse scenarios, but such approaches could require an excessive computational effort if different load combinations are to be analysed. Later a theory of structural vulnerability is reviewed (section 6) which quantifies alternate load paths without the magnitude and location of loads. A recent National Institute of Standards and Technology (NIST) report[18] gives practical measures to reduce the chances of progressive collapse of different types of buildings.

5. ENERGY-BASED APPROACHES TO ROBUSTNESS

The capacity to absorb energy depends on the ductile behaviour of the materials in a structure. A quantitative measure of the energy absorption capacity per unit volume of a material is given by the area under the stress–strain curve. Based on this principle, Smith[19] proposed an energy approach to determine the sequence of members which, if failed, would result in the collapse of a structure with the least amount of energy required. In this method the balance of energy in each member at each damaging event is considered. Firstly, the work of failure W_f of each member under different loading conditions (i.e. buckling, bending, tension, etc.) is determined. This uses the force–displacement curve. Secondly, the structure is subjected to a specified set of loads, and energy ratios are calculated for all the members. The energy ratio is defined as the ratio of the difference between the work of failure and the strain energy in each member to the sum of the strain energy in all members. If a member is loaded in tension, then the work of failure in tension is used. The member with the smallest energy ratio is selected and considered as failed by notionally removing it from the structure. The damaged structure is then analysed, the process is repeated and another member is removed. This process continues until the structure becomes a mechanism. The final result is the sequence of failed members which requires the least amount of energy to collapse the structure.

Beeby[20] proposed a method using the amount of energy absorbed before collapse. The energy absorption capacity of a structure was defined as the product of the volume of the material forming the structure and a coefficient which is derived from the strain energy at failure and the type of material. This is applied to any possible mode of failure and the minimum amount of energy required to cause collapse is used as a measure of robustness. Although this method does not include the influence of redundancy, it highlights the importance of the energy absorption capacity of a structure as a key factor to achieve robustness.

Another approach which examines the stability of a damaged structure using energy principles was proposed by Yankelevsky et al.[21] The stability of the structure under four predefined failure modes is determined by the ratio of the work done by the external gravitational forces and the internal resistive forces. This approach is similar to Smith[19] mentioned previously, in that it identifies (a) the failure mode that requires the least amount of energy to collapse the structure and (b) possible paths through which damage will propagate. An experimental approach to evaluate the energy required to damage a structure during an earthquake was given by Akiyama.[22] The energy producing damage in the structure is calculated as the difference between the total seismic energy input and the energy absorbed by damping. An important aspect of this method is that it is applied to the whole structure including the connections. The total energy input can be calculated for any type of loading, thus, its application can be extended to include accidental loads. Among other studies, Dusenberry and Hamburger[23] have suggested energy-based approaches which capture the potential for collapse, and Zhou and Yu[24] have shown the use of energy absorbing devices to arrest progressive collapse.

6. STRUCTURAL VULNERABILITY THEORY

An insight into the lack of robustness is gained by identifying how a system is vulnerable, as this indicates how it is weakest. A structure that is not properly configured or formed is a potential hazard. Progressive collapse is often analysed as a property of the response of the structure to loading. That response emerges from the physical form of the structure, however, that is the way the structure is connected. The magnitude of the loading is important to determine the quantitative parameters of the response, but the qualitative nature of the response derives from the structural form. Considering the nature of threat owing to unforeseen hazards and that resulting from extreme events, significant protection against disproportionate collapse can be achieved by improving the form of a structure.

Structural vulnerability theory[25-27] helps to achieve in-built robustness. The theory enables the form of a structure to be described so that the quality of its connectivity can be measured. This quality is called the 'well-formedness' of the structure. A measure of well-formedness, based on the stiffness matrices, is used to create a hierarchical model of the structure. The hierarchical model starts with the identification of primitive structural 'rings' made of joints and members at the first and familiar definition of a structure. A structural 'ring' (in 2D) or a 'round' (in 3D) is a load path which, by its configuration, is capable of resisting an arbitrary set of applied forces. Clusters of these rings are then formed according to their well-formedness and the degree of connectivity. New sets of rings of clusters are formed to provide a second level of definition of the structure. The process of clustering is repeated to form even higher levels of definitions until one single cluster, the whole structure, remains. These clusters are then 'unzipped' by a series of deteriorating events to form a failure scenario. The most vulnerable failure scenario is the one where the least damage causes the maximum consequences. Consequences are measured in terms of separation from the supports. The minimum demand failure scenario shows how a structure can be damaged with the least amount of effort. The analysis concludes by finding different types of failure scenarios and which include the minimum demand scenario and the most vulnerable scenario.

In more recent work,[28] the failure scenarios obtained using this theory have been examined for specific loading condition and

a new measure of hazard potential has been proposed. This measure gives the potential for the progression of damage through a particular failure scenario under a stated set of loads. The measure depends on the flow of potential energy of the loads into internal strain energy caused by the damage event and the consequent change in well-formedness. This potential for energy flow is the hazard potential, which is analogous to temperature in thermodynamic systems. The hazard potential of a structure before a particular damage event is defined as the ratio of the change in strain energy to the change in well-formedness owing to that event. As damage progresses the hazard potential increases until it reaches a maximum. This is a critical state since further damage by a trigger event will cause the structure to collapse totally or progressively. For a set of failure scenarios in a structure, the lowest hazard potential denotes the most vulnerable failure scenario to the assumed set of loads because it requires the least effort to cause maximum consequences. For different structures the least robust structure is the one that has the highest hazard potential for the release of energy when failed in its most vulnerable way, as it has the highest potential for consequential damage.

Vulnerability is different from reliability, but they are both a part of risk. Structural reliability theory is concerned with calculating the probability of a demand exceeding the capacity of a system. The modelling takes into account the uncertainties of the variables and determines the probability of failure to a specific loading condition. The results are normally used in the quantitative risk analysis of a structure but low probability failure scenarios, which result in high consequences, can be missed.

7. MANAGING RISK

UK building regulations[11] require a systematic risk assessment for class 3 buildings. There is a similar requirement in EN 1990[13] where a qualitative risk analysis is suggested as a minimum, and where relevant and practicable, a quantitative risk analysis should also be carried out. For qualitative risk analysis, all hazards and hazard scenarios should be identified and the consequences for structural safety should the structure fail to withstand the hazards, should be determined. For this purpose, a variety of techniques such as fault tree, event tree and so on are suggested. Quantitative risk analysis uses probabilistic estimates for all undesired events and risk is determined as the mathematical expectation of the consequences of the undesired events. Equation (1)[14] is suggested for this purpose

$$R = \sum_{i=1}^{N_H} p(H_i) \sum_{j}^{N_D} \sum_{k=1}^{N_S} p(D_j|H_i)\, p(S_k|D_j)\, C(S_k) \qquad 1$$

where
N_H is the number of hazards
N_D is the number of damage states owing to hazards
N_S is the number of adverse states following the damage
$p(H_i)$ is the probability of occurrence of the ith hazard
$p(D_j|H_i)$ is the conditional probability of the jth damage state given hazard i
$p(S_k|D_j)$ is the conditional probability of the kth adverse state given damage j
$C(S_k)$ is the consequence owing to adverse state S_k.

This is a useful model but it is not easy to obtain these probabilities owing to the nature of hazards and dependency between the events. Also, the measures of consequences may vary. Ellingwood and Dusenberry[29] provide a good discussion on the above probabilities.

If only the structural consequences are considered, consequences C for a given damage D will be a function of the inherent form of the structure. This has been defined as vulnerability in the previous section and hence it is possible to determine structural risk as an additional step to vulnerability analysis. The hazards that can cause these failure scenarios need to be identified and included in the analysis. A methodology to calculate the structural risk of vulnerable failure scenarios has been suggested by Agarwal.[30] It uses a risk matrix in terms of the likelihood of failure scenarios and the vulnerability to determine the degree of risk. The results can be used to identify actions which may result in the progressive collapse of a structure and an efficient means for counteracting them. The approach also avoids the need to run a number of numerical simulations either by 'notionally' removing a member or with different load patterns. Thus it could become a significant tool in meeting the regulatory requirements.

The above approach has parallels with the catastrophe modelling approach[31] used within the insurance industry to quantify the risk posed by the natural hazards. Of the three steps—hazard modelling, vulnerability modelling and financial modelling—the first two are particularly relevant. As part of hazard modelling, a catalogue of stochastic events is simulated and each event is associated with a magnitude and frequency depending on the geographic location. Then vulnerability curves are developed using numerical engineering methods. These curves relate hazard intensity and mean damage ratio[31] for a structure. Loss data from historic events are also used to calibrate these curves. In vulnerability analysis, loss of well-formedness is equivalent to damage ratio. The relationship between this ratio and the intensity of the loads results in a vulnerability curve.[32] These curves can be produced for various characteristic buildings and for different failure scenarios. These curves cannot, however, be calibrated because of the paucity of historical data on progressive failures.

The biggest challenge in undertaking any risk assessment is the identification of hazards. Hazards with an extremely small chance of occurrence but with a potential for high consequences require special attention. The robustness of a structure should be considered in the context of such low-probability high-consequence events rather than the high-probability events alone.

Recently, Scoss has proposed a risk-based holistic approach[33] to the process of ensuring robustness where appropriate people, processes and products are utilised throughout the life cycle of a project. Robustness is considered as an aspect of quality. Hence, it is necessary that the people involved in the design, construction and subsequent management of a structure are competent. They need to follow best practice and monitor the processes not just for analysis and design but also for communication and coordination. A process model, proposed by Blockley and published in this special issue,[34] could be useful in monitoring such issues and hence managing risk.

8. CONCLUSIONS

The structural engineering community is making an increasing effort to ensure that disproportionate collapse is avoided and that structures are sufficiently robust. There is a broad agreement on the methods to examine the potential for progressive collapse. This generally requires a non-linear static/dynamic of a structure under specified design loads after the notional removal of one or more members. Although such an analysis helps to improve the robustness of a structure, there is no satisfactory measure of robustness. Indirect measures based on the expected consequences if damage were to occur, are used to achieve robustness.

The form of a structure is the most important factor in achieving a robust structure under any loading, especially ones that may be unforeseen. Structural vulnerability analysis can help identify the weaknesses in the form of a structure. These are given by vulnerable failure scenarios, which must be avoided if the risk these pose is found to be high. Appropriate measures should be taken to manage the risks. A robust structure will have low risk but the converse is not necessarily the case.

REFERENCES

1. BLOCKLEY D. I. *The New Dictionary of Civil Engineering.* Penguin, London, 2005.
2. GROSS J. L. and MCGUIRE W. Progressive collapse resistant design. *ASCE Journal of Structural Engineering*, 1983, **109**, No. 1, 1–15.
3. MALLA R. B. and NALLURI B. B. Dynamic nonlinear member failure propagation in truss structures. *Structural Engineering and Mechanics*, 2000, **9**, No. 2, 111–126.
4. MARJANISHVILI S. M. Progressive analysis procedure for progressive collapse. *Journal of Performance of Constructed Facilities*, 2004, **18**, No. 2, 79–85.
5. KAEWKULCHAI G. and WILLIAMSON E. B. Beam element formulation and solution procedure for dynamic progressive collapse analysis. *Computers and Structures*, 2004, **82**, Nos 7–8, 639–651.
6. GRIERSON D. E., XU L. and LIU Y. Progressive-failure analysis of buildings subjected to abnormal loading. *Computer-Aided Civil and Infrastructure Engineering*, 2005, **20**, No. 3, 155–171.
7. WARN G., BERMAN J., WHITTAKER A. and BRUNEAU M. Reconnaissance and preliminary assessment of a damaged high-rise building near ground zero. *Structural Design of Tall and Special Buildings*, 2003, **12**, No. 5, 371–391.
8. INSTITUTION OF STRUCTURAL ENGINEERS. *Safety in Tall Buildings*. IStructE, London, 2002.
9. MEAS M. A., FRITZSONS K. E. and GLOWIENKA S. Structural robustness in the light of risk and consequence analysis. *IABSE Structural Engineering International*, 2006, **16**, No. 2, 108–112.
10. STANDING COMMITTEE ON STRUCTURAL SAFETY. *Structural Safety 1994–96 Review and Recommendations*. SCOSS, London, 1997. Available online at: http://www.scoss.org.uk/publications/rtf/11Report.pdf.
11. OFFICE OF DEPUTY PRIME MINISTER. *Building Regulations 2000: Approved Document A: Structure*. ODPM, London,

2004. Available online at http://
www.planningportal.gov.uk/england/professionals/en/
4000000000067.html.

12. BRITISH STANDARDS INSTITUTE. *BS 5950: Part 1. Structural Use of Steelwork in Building. Code of Practice for Design in Simple and Continuous Construction*. BSI, London, 2000.

13. EUROPEAN COMMITTEE FOR STANDARDISATION. *Eurocode: Basis of Structural Design*. Comité Européen de Normalisation, Brussels, 2002. EN 1990.

14. EUROPEAN COMMITTEE FOR STANDARDISATION. *Eurocode 1: Actions on Structures. Part 1-7: General actions– accidental actions*. Comité Européen de Normalisation, Brussels, 2006. EN 1991-1-7.

15. AMERICAN SOCIETY OF CIVIL ENGINEERS. *Minimum Design Loads for Buildings and Other Structures*. ASCE, Reston, VA, 2005. ASCE-7.

16. US GENERAL SERVICES ADMINISTRATION. *Progressive Collapse Analysis and Design Guidelines for New Federal Office Buildings and Major Modernisation Projects*. GSA, Auburn, Washington, 2003. Available online from htpp:// www.oca.gsa.gov.

17. US DEPARTMENT OF DEFENCE. *Unified Facilities Criteria: Design of Buildings to Resist Progressive Collapse*. DoD, Washington, 2005. UFC 4-023-03. Available online from htpp://www.wbdg.org/ccb/.

18. NATIONAL INSTITUTE OF STANDARDS AND TECHNOLOGY. *Best Practices for Reducing the Potential for Progressive Collapse in Buildings*. NISTR, Gaithersburg, MD, 2007. NISTR 7396.

19. SMITH J. W. Energy approach to the assessment of corrosion damaged structures. *Proceedings of the Institution of Civil Engineers, Structures and Buildings*, 2003, 156, No. 2, 121–130.

20. BEEBY A. Safety of structures and a new approach to robustness. *Structural Engineer*, 1999, 77, No. 4, 16–21.

21. YANKELEVSKY D. Z., YAGUST V. I., DANACYGLER A. N. and SCHWARZ S. Final state of damage in buildings subjected to extreme loading. *Response of Structures to Extreme Loading, Proceedings of XL2003, Toronto*, 2003.

22. AKIYAMA H. Evaluation of fractural mode of failure in steel structure following Kobe lessons. *Constructional Steel Research*, 2000, 55, Nos 1–3, 211–237.

23. DUSENBERRY D. O. and HAMBURGER R. O. Practical means for energy based analysis of disproportionate collapse potential. *Journal of Performance of Constructed Facilities*, 2006, 20, No. 4, 336–348.

24. ZHOU Q. and YU T. Use of high-efficiency energy absorbing device to arrest progressive collapse of tall building. *ASCE Journal of Engineering Mechanics*, 2004, 130, No. 10, 1177–1187.

25. WU X., BLOCKLEY D. I. and WOODMAN N. J. Vulnerability of structural systems. Part1: rings and clusters. Part 2: failure scenarios. *Journal of Civil Engineering Systems*, 1993, 10, No. 4, 301–333.

26. LU Z., YU Y., WOODMAN N. J. and BLOCKLEY D. I. Theory of structural vulnerability. *Structural Engineer*, 1999, 77, No. 18, 17–24.

27. AGARWAL J., BLOCKLEY D. I. and WOODMAN N. J. Vulnerability of 3-dimensional trusses. *Structural Safety*, 2001, 23, No 3, 203–220.

28. ENGLAND J. C., AGARWAL J. and BLOCKLEY D. The vulnerability of structures to unforeseen events. *Computers and Structures*, 2008, 86, No. 10, 1042–1051.

29. ELLINGWOOD B. R. and DUSENBERRY D. O. Building design for abnormal loads and progressive collapse. *Computer-Aided Civil and Infrastructure Engineering*, 2005, 20, No. 3, 194–205.

30. AGARWAL J. Managing structural risks. *Proceedings of ICOSSAR 2005, Safety and Reliability of Engineering Systems and Structures* (AUGUSTI G., SCHUELLER G. I. and CIAMPOLI M. (eds)). Millpress, Rotterdam, 2005, pp. 1793–1799.

31. GROSSI P. and KUNREUTHER H. *Catastrophe Modelling: A New Approach to Managing Risk*. Kluwer, New York, 2005.

32. ENGLAND J. C. *Structural Vulnerability and Hazard Potential*. PhD thesis, University of Bristol, Bristol, 2005.

33. STANDING COMMITTEE ON STRUCTURAL SAFETY. *A Risk Managed Framework for Ensuring Robustness*. SCOSS, London, 2006. Paper No. SC/06/28. Available online from http://www.scoss.org.uk/publications/rtf/ Risk_Framework_paper.pdf.

34. BLOCKLEY D. Managing risks to structures. *Proceedings of the Institution of Civil Engineers, Structures and Buildings*, 2008, 161, No. 4, 231–237.

What do you think?

To comment on this paper, please email up to 500 words to the editor at journals@ice.org.uk

Proceedings journals rely entirely on contributions sent in by civil engineers and related professionals, academics and students. Papers should be 2000–5000 words long, with adequate illustrations and references. Please visit www.thomastelford.com/journals for author guidelines and further details..

Proceedings of the Institution of
Civil Engineers
Structures & Buildings 161
August 2008 Issue SB4
Pages 189–197
doi: 10.1680/stbu.2008.161.4.189

Paper 700044
Received 06/07/2007
Acccepted 04/03/2008

Keywords: buildings, structures &
design

Mauricio Sánchez-Silva
Associate Professor,
Department of Civil
Engineering, Universidad de
Los Andes, Bogota,
Colombia

David V. Rosowsky
A. P. & Florence Wiley Chair,
Professor, Department of
Civil Engineering, Texas A&M
University, USA

Risk, reliability and sustainability in the developing world

M. Sánchez-Silva PhD and D. V. Rosowsky PhD, PE

Developing countries need more affordable housing and a plan for infrastructure development that is environmentally sound, socially acceptable, economically justified and sustainable. The enormous shortage of funds as well as pressures resulting from other factors (e.g. poverty) can, however, make long-term and sustainable planning exceedingly difficult. A new focus should be placed on developing models that are able to use and process larger amounts of information, obtained from different sources and in different forms, to provide engineering solutions that have a positive impact on sustainability. Achieving sustainable development includes taking risks and using limited resources efficiently. Towards this end, structural risk and reliability are essential tools for defining appropriate levels of investment. It is suggested that, by adopting safety standards specified in codes of practice of developed countries, less developed countries may be spending more to achieve levels of safety not warranted by their level of development. That is, it may be more beneficial to accept a lower level of safety in favour of increased infrastructure development.

1. INTRODUCTION

Sustainable development refers to the continued socio-economic growth by the rational use of natural resources and the appropriate management of the environment. It is not a fixed goal, but rather is a continuous and long-term commitment. For developing countries, the World Bank has set specific targets encompassing access to safe drinking water and basic sanitation and improvements in the conditions of the lives of slum dwellers.[1] Housing and infrastructure needs in developing countries are substantial. These countries need an infrastructure development that is environmentally sound and socially acceptable as well as financially sustainable. The shortage of funds and pressures from other factors (e.g. poverty), however, present challenges to long-term and sustainable planning.[2]

Risk is concerned with what *might* happen, with identifying future scenarios and assessing their chances of occurrence within a given context. Risk analysis is a tool for decision making; decisions leading to sustainable development therefore involve taking risks. The pace at which policies and decisions are transformed into improving long-term life quality with intergenerational responsibility depends upon the risks taken.

As such, the relationship between risk and sustainability cannot be restricted to a probabilistic approach since it is conditioned by context (e.g. socio-economic and political issues). The same basic data and information in two different environments may lead to different assessments and conclusions: in other words, different decisions. Sustainable development must be based upon strategic risk-based decisions.

The paper is aimed at

(a) reviewing critically the traditional view of risk and reliability within the context of sustainable structural design and the need for a broader approach
(b) discussing the implications of measurable indicators of sustainability on structural design, within different socio-economic climates
(c) describing the implications of structural optimisation of design parameters on sustainability.

2. DECISION-MAKING CONSIDERATIONS IN ENGINEERING DESIGN

Appropriate engineering decisions depend upon the quality and quantity of the evidence available and the relevance of models for the problem at hand. Any engineering decision is therefore made within a space defined by: the scope and nature of the decision; the amount and quality of information; and the complexity of models (Fig. 1).

Regarding the scope and nature of decisions (Fig. 1), classical engineering design of large infrastructure systems has been advanced recently by including in the analysis economic considerations and the structure's expected life-cycle performance. The former has yielded the so-called economic-based engineering, where decisions include not only safety and functionality but also the effectiveness of investment. The use of life-cycle approaches attempts to take into consideration the context of the project, more complex socio-economic considerations and the societal impact of the decisions.[3,4] Decisions about sustainability have been controlled by socio-economic and political pressures with engineering models playing only a minor role.

Long-term sustainable decisions can only be made based on information that describes different sides of the problem—that is, technical aspects as well as the social, economic and political issues. Nowadays, the capacity for collecting and

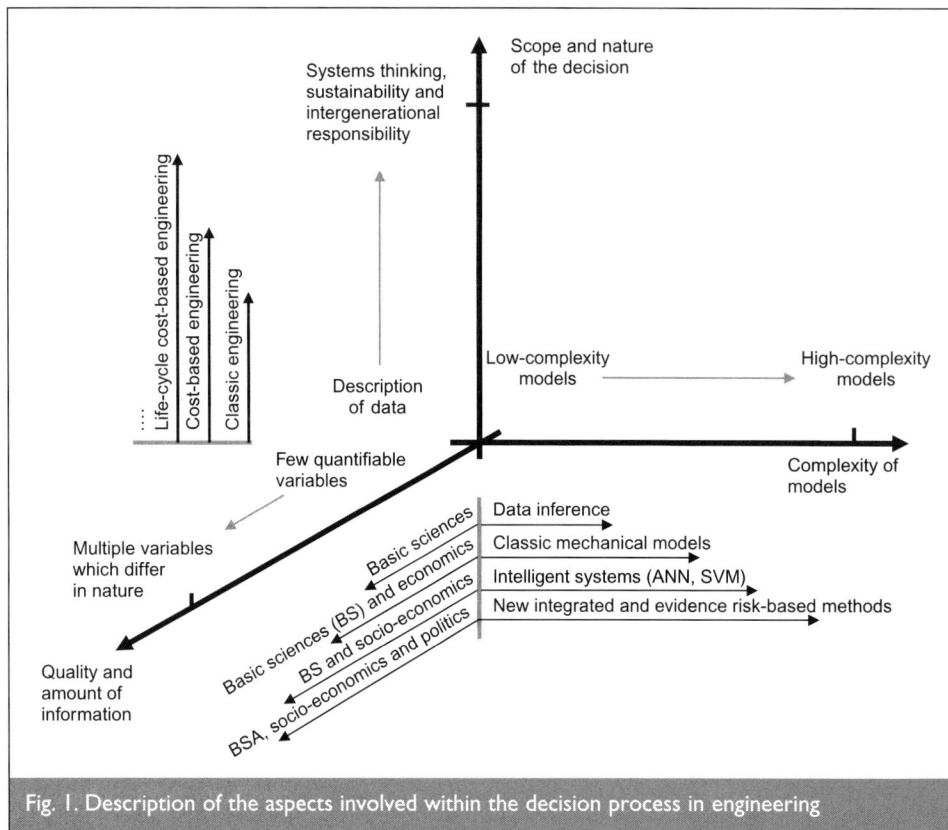

Fig. 1. Description of the aspects involved within the decision process in engineering

computational capabilities such as neural networks, support vector machines and biology-based techniques, which have been developed to address highly specific technical problems. The complexity of decisions that actually impact society (i.e. sustainability), however, make existing tools for risk analysis appear incomplete. New tools, which are able to integrate existing models with evidence coming from different sources (e.g. social, economic and political aspects) must be developed.

Figure 2 depicts the state of current engineering practice and the direction of future development. Traditional engineering disciplines should expand in all directions (i.e. dotted region in Fig. 2). This implies encompassing the interests of larger sectors of society by developing models able to use and process larger amounts of information, obtained from different sources, to provide engineering solutions with a positive impact on sustainability.

processing large amounts of physical data is almost unlimited. The contribution of more data does not, however, necessarily provide additional relevant information for making effective decisions. Other factors, which differ in nature and which are not as easy to quantify, should also be taken into account. For instance, economic aspects include interest rates, behaviour of local and regional markets, economic growth, inflation and so on; social considerations include income distribution, equality issues, employment and poverty; and political issues should consider national strategic decisions, development goals, regional interests and so on. Although these aspects may seem distant from engineering practice, they are closely related to structural engineering—in particular, in mid-size and small economies such as those of most developing countries. For instance, Sánchez-Silva and Arroyo[5] showed that reviewing and re-evaluating target reliabilities defined in the code of practice in light of local risk acceptance criteria could lead to significant economic savings and the possibility of providing the poorest sectors of society with low-income housing and improved life quality. In summary, sustainability-related decisions are usually made based on simplified technical models which use a small amount of information (evidence) from a large numbers of sources in an attempt to provide better decision tools.[6]

The final aspect in the decision-making process is related to the need to make better predictions under uncertainty. Classic approaches are based mostly on experimental data or information extracted from field observations. During the past 20 years, the basic reliability problem (i.e. basic resistance–demand problem) and corresponding solution techniques (e.g. first- and second-order reliability models, i.e. FORM, SORM) have evolved and have been widely implemented in many engineering contexts. Recently, data inference and classic mechanical models have been revised in light of new

3. SUSTAINABILITY

3.1. Definition and measurement
Among existing interpretations, the Brundtland Commission[7] provides a widely accepted definition of sustainable development: 'Meeting the needs of the present without compromising the ability of future generations to meet their needs.' Sustainability is a concept that has been discussed in many contexts and, depending upon the discipline, applied to specific topics (e.g. economics and biology). In the current paper, and consistent with the *Agenda 21 for Sustainable Construction in Developing Countries*,[8] sustainable construction is defined as '. . .a holistic process aiming to restore and maintain harmony between the natural and built environments, and create settlements that affirm human dignity and encourage economic equity.' Within this context, sustainable development should seek to provide people with opportunities for an acceptable quality of life by protecting the physical environment and its resources. It has been argued in various forums that current concerns about sustainability have become a new ethical standard related to intergenerational equity with implications for the development of civil infrastructure. These implications have to do with aspects such as pollution control, rational use of resources and financial feasibility of engineering projects.

3.2. Sustainability and construction in developing countries
The construction industry is important to the economy and sustainability of any country. It constitutes approximately

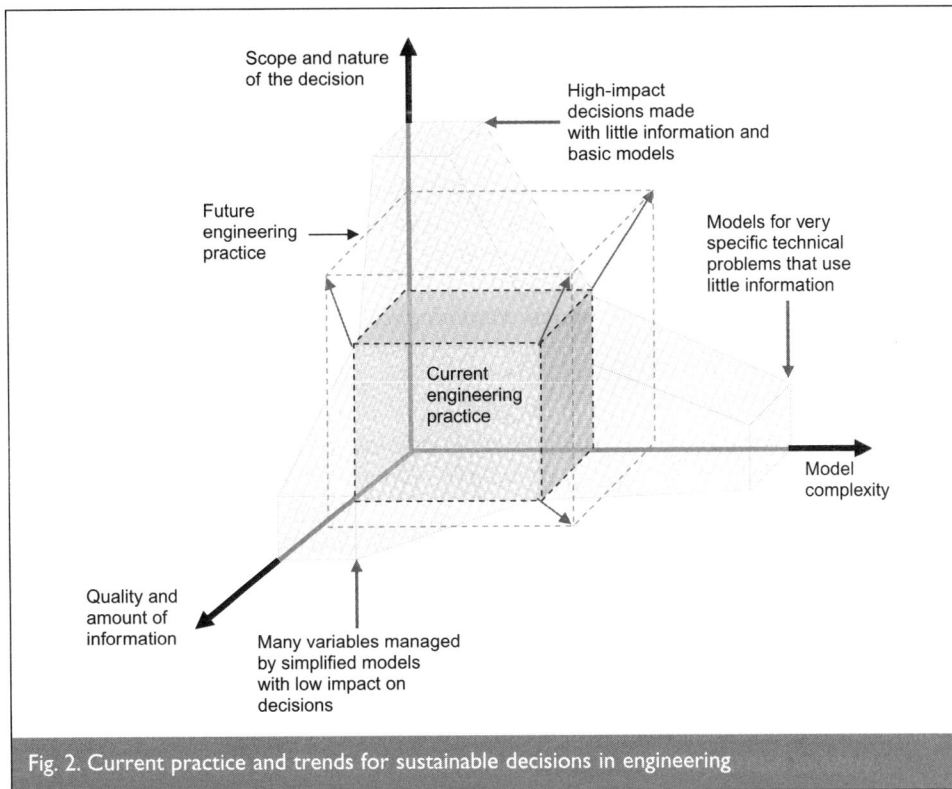

Fig. 2. Current practice and trends for sustainable decisions in engineering

Within the figure:
- Scope and nature of the decision
- High-impact decisions made with little information and basic models
- Future engineering practice
- Models for very specific technical problems that use little information
- Current engineering practice
- Model complexity
- Quality and amount of information
- Many variables managed by simplified models with low impact on decisions

energy-intensive materials and the production of both iron and steel is responsible for 4·1% of global energy use.[13]

The social aspect of sustainability relates to the impact of the construction industry on life quality. The construction industry is the largest industrial employer in the world with 111 million employees worldwide, of which 74% are in low-income countries.[14] Labour in the construction industry often comes from the poorest sector of society, which is characteristic of many developing countries. Balancing short- and long-term needs is a challenge for sustainability in the developing world, where short- and medium-term needs are much more urgent. If this compromise is not achieved, people living in poverty have little option but to meet only their short-term needs. Given the short-term focus of people in poverty, consideration of longer-term environmental goals will need government support.[11] According to The International Council for Research and Innovation in Building and Construction (CIB) and United Nations Environment Programme International Environmental Technology Centre (UNEP-IETC)[8] sustainable construction in developing countries is feasible only if it meets the needs and requirements of the people and does not conflict with their culture and values.

In almost every country in the world, the built environment constitutes more than half of the total national capital investment, and construction represents as much as 10% of the GDP.[15] Poorly designed construction that incurs high operating costs over its life will be an economic as well as an environmental burden for developing countries. An economically efficient construction industry enhances environmental sustainability by ensuring least-cost methods of construction that encourage robust allocation of resources and discourage waste. Furthermore, economic sustainability within construction requires that social and environmental costs are internalised and reflected in the final product prices.[8] For developing countries, the objectives of future sustainability are bound together with the need to attain a sustainable present grounded in economic growth.

3.3. Codes of practice and sustainability
Recent reports from international organisations make evident that the construction industry plays a larger role in developing countries than any other sectors. This is not the case in developed countries where technology and services, among others, are the main contributors to economy. The World Bank[16] states that: 'improving infrastructure in developing countries is increasingly recognised as a key factor in reducing

8–12% of the gross domestic product (GDP) in many developing and industrialised economies.[9] If large infrastructure projects are included, this figure can reach or exceed 20%. United Nations (UN) estimates of housing needs over the coming decades indicate that the number of households in the less developed regions of the world will double every 15–20 years. The estimated deficit in 2000 was 1108·3 million households and the predictions for 2015 and 2020 are 1583 and 2074 million respectively.[10] Statistics show that between 2015 and 2020 the less developed regions of the world will require ten times the housing of the more developed regions. These figures are indicative of the challenges for developing countries. The impact of the construction industry on sustainability will be significant and can be grouped into environmental, social and economic.

The environmental impact of the construction industry is larger in developing countries because they are virtually always under construction with a low degree of industrialisation. The impact is a consequence of resource consumption and environmental degradation. First, the construction sector is linked with the extensive use of natural resources owing to the rapid growth in global population and housing demands. Second, the impact of the construction industry on the environment is related to consumption of energy and greenhouse gas emissions.[11] Environmental pollution is caused by the processing of the raw materials and manufacturing of cement and concrete, which causes pollution of watercourses, atmospheric pollution, deforestation and soil erosion. To this can be added, manufacturing processes of materials that form the basis of modern construction; for example, concrete and steel are some of the biggest contributors to climate change. Cement production is, after the burning of fossil fuels, the biggest anthropogenic contributor to greenhouse gas emissions.[12] On the other hand, steel is one of the most poverty, increasing growth and reaching the Millennium Development Goals (MDGs).' The impact of engineering decisions, related to the design and operation of structural systems, on the sustainability of any country is mostly defined by codes of practice. In structural engineering, risk levels are hidden within code provisions and therefore many decisions about safety, functionality, feasibility and the long-term impact on society have been made before the project is conceived. Evidence has shown that there is a correlation between the level of social development and risk acceptance. As society becomes more developed, risk tolerance decreases. These differences, however, have not been captured in structural design and construction. Most developing countries have adopted design standards established by developed countries (e.g. Eurocode–Comité Européen de Normalisation (CEN), International Building Code). This implies that they use safety and reliability requirements that are not necessarily in accordance with their own socio-economic conditions. These decisions severely affect their development by not taking into account basic criteria of risk acceptability and unnecessarily increasing costs.

4. RELIABILITY-BASED OPTIMISATION OF STRUCTURES

4.1. Cost-benefit objective function

A significant amount of research has been conducted in the area of economic-based optimisation of structural design parameters taking into account the project life cycle. Life-cycle analysis provides a framework for considering design and operational costs over the entire life of a project.[17,18] The life cycle includes production, operation, maintenance, deconstruction and resource recovery. The objective function that describes the performance of an engineering system (in financial terms) throughout its life cycle can be expressed as

$$1 \quad Z(p) = B(p) - [C(p) + M(p) + F(p) + D(p)]$$

where $B(p)$ are the benefits; $C(p)$ are the planning, design and construction costs; $M(p)$ are the inspection, maintenance and operation costs; $F(p)$ are the financial and economic costs; and $D(p)$ the failure costs. The vector parameter p includes all variables that describe the performance of the system (i.e. physical, operational and financial). Several closed-form solutions for equation (1) have been proposed, usually by describing the project performance as a renewal process and assuming a continuous and constant discount rate. For further details see, for example, Rackwitz et al.[19] and Wen and Kang.[20] Finding a comprehensive and unique analytical formulation is not, however, an easy task since analytical formulations are subject to strong assumptions about the discount rate and other financial factors, structural performance of the system, uncertainty models and inspection and maintenance policies, among others.

4.2. Risk-balanced structural design

For the project to be feasible, discounted costs at time $t = 0$ in equation (1) should be lower than benefits. When an analytical solution is available, the selection of the optimal design parameter (e.g. peak ground acceleration) is obtained by maximising the expected benefit or minimising the expected

life-cycle cost. This approach is adequate only if the decision maker is risk-neutral. When this is not the case, there are tools such as expected utility theory that can take into account the preferences of decision makers. In addition, for large and complex problems, it has been observed that in some cases near the optimal design point the slope of the expected life-cycle cost curve is relatively flat. This implies that finding a clear minimum expected life-cycle cost (or maximum benefit) is not straightforward, considering the usually large uncertainty of the variables. In these cases, one option is to use mathematical tools such as stochastic dominance criteria[21] to find a set of optimal reasonable solutions instead of a unique solution.[22] These inconclusive results may, however, be a consequence of the complex nature of the factors involved (e.g. quantification of the context). Then, in complex problems, the term optimisation in the mathematical sense is quite restrictive since 'optimal' decisions are context dependent and optimised systems are still vulnerable to unforeseen events. As the decisions become controlled by aspects that cannot be modelled by using standard numerical techniques, the term optimum design should be replaced by risk-balance design.

4.3. Sustainable optimum structural decisions

A numerical model of the structure's life cycle (e.g. equation (1)) requires the selection of parameters that may actually be useful to support sustainable decisions. Along these lines, the global reporting initiative groups sustainable performance indicators (SPIs) into the following three categories: economic, social and environmental (see also United Nations Conference on Environment and Development (UNCED)[23]). The first group of indicators focuses on how the economic status of the stakeholders (i.e. society) changes as a consequence of the organisation's activities (i.e. construction industry). Social performance indicators describe the impact that an organisation might have on the social system within which it operates, for example labour practices, consumer habits and human rights. Finally, environmental indicators describe the impacts on natural systems including ecosystems, land, air and water.[24] Many reports have suggested a large number of SPIs in an attempt to cover the wide and diverse nature of sustainability (see International Sustainable Indicator Network (ISIN)[25]). Nevertheless, the need to achieve a compromise between the practical issues and a broader view in structural design has led to the selection of the indicators shown in Table 1. The selected indicators are in agreement with those selected by the UN for the development of the human development index (HDI).[26] Although environmental aspects are important and should be included as part of any comprehensive analysis, they will not be considered in the present paper since their quantification is still a matter of great debate.

Any model related to sustainability must address the problem of health and life quality. Risk to life should not be restricted to the possibility of dying. Based on the work of Nathwani et al.,[27] the societal life quality index (SLQI) has been proposed as a public measure of what society can afford for an anonymous member in terms of risk to life under any threatening event. The SLQI takes into account life expectancy, consumption and time devoted for an individual to raise his income.[19] Economic growth is commonly measured by the annual variation of GDP, which is a macro-economic indicator that captures the average well-being of society. Statistics show that societies with large

Description	SPI group	Indicator
Health and long life	Social	Societal life quality index
Economic growth	Social/economic	GDP per capita
Financial stability	Economic	Interest rates
Long-term policy and regulation	Socio-economic and political	Average real exchange rate variation

Table 1. Main factors and indicators contributing to sustainable development

GDP/capita are wealthier. Long-term economic growth depends upon many factors but is usually attached to financial stability; for example, small variations in interest rates, low and controlled inflation and stable exchange rate. While economic growth describes the overall well-being of society, financial stability provides the means actually to make this growth possible. Political issues and regulation provide the context within which political stability and economic growth take place. Although it is difficult to find a unique indicator that actually encompasses all dimensions of the socio-political stability of a country, evidence has shown that countries with frequent and strong social and political changes have a lower or very unstable economic growth. This is true mainly because the financial conditions are not conducive to private investments. Although the discussion in the literature is vast

and is the subject of considerable controversy, there is evidence that average real exchange rate variation is a factor that provides a useful indicator of this dimension. As stated by Carmignani et al.:[28] 'In choosing the exchange rate regime, policymakers are concerned with its future sustainability.'

As an example of the impact of infrastructure on sustainability, Fig. 3 shows the relationship between km of paved roads with all four indicators of sustainability shown in Table 1. 'Km of paved roads' has been used as a surrogate measure of infrastructure development and the level of activity in the construction industry. It is observed that selected indicators improve as the number of kilometres of roadways increase. In other words, the construction of infrastructure leads to higher sustainability indicators.

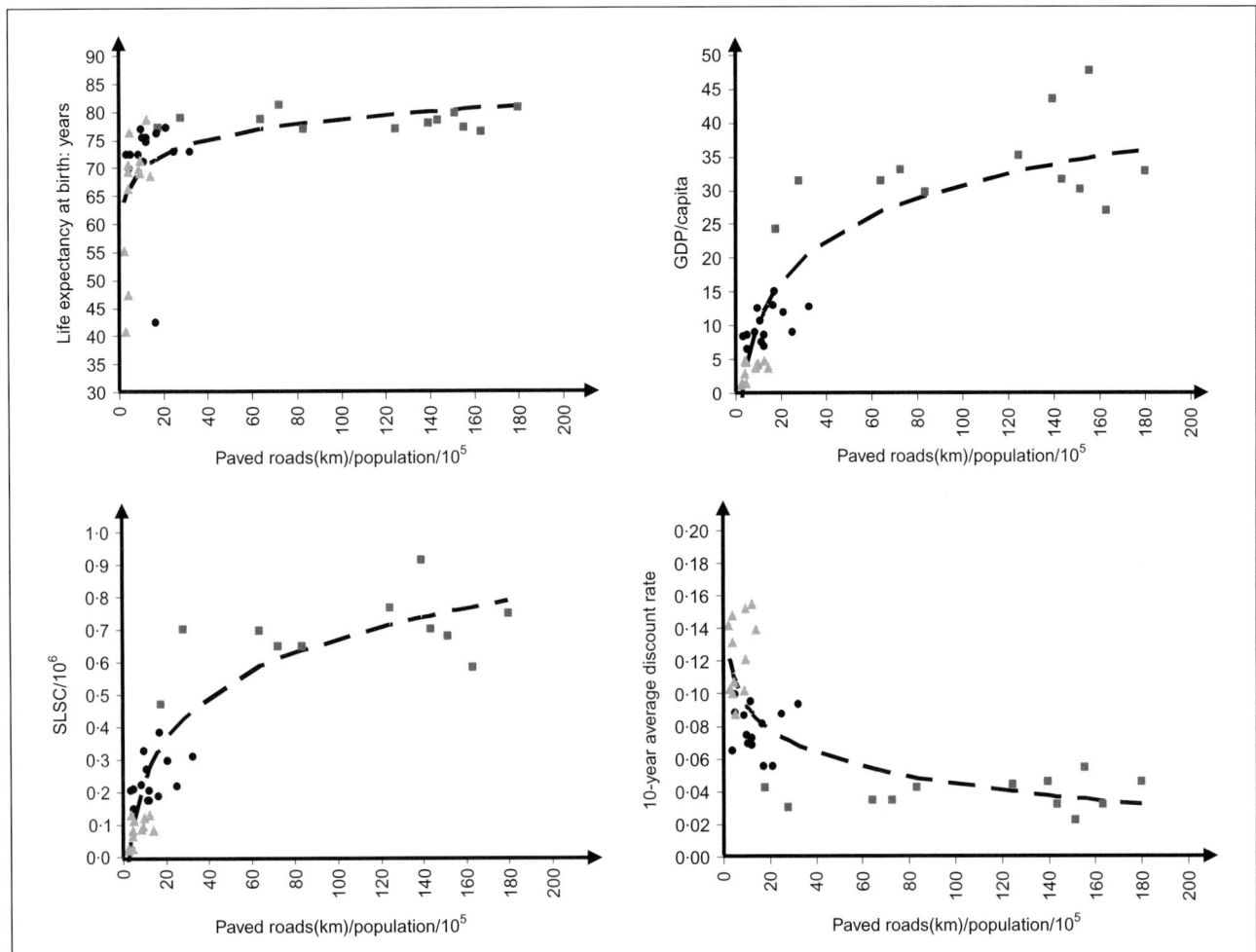

Fig. 3. Correlation between transport infrastructure and selected SPI for 37 selected countries (Table 2) (Underdeveloped, developing and developed countries). Data sources: Sanchez and Rackwitz,[29] Rackwitz,[30] CIA,[31] Deutsche Bank[32] and World Bank[33]

5. EFFECT OF SUSTAINABILITY INDICATORS ON THE OPTIMAL STRUCTURAL DESIGN

5.1. Design criteria by optimisation

This section illustrates how selected SPIs (Table 1) affect the seismic design of structural systems in different socio-economic contexts. The results for 37 countries, divided into three levels of development, are shown in Table 2 and Fig. 4. The seismicity is assumed to be the same in all cases and is defined by a stationary Poisson process with random independent earthquake intensities and occurrence rate $\lambda = 0.05$ ($1/\lambda = 20$ years). It is expected that if a structure fails, it will be reconstructed systematically after failure. Based on these considerations, the design criteria can be obtained by maximising the following objective function[3]

$$2 \qquad Z(p) = \frac{b}{\gamma_c} - C(p) - [C(p) + H]\frac{\lambda\, p_f(p)}{\gamma_c}$$

where p is the design criteria; b is the benefit (i.e. % of construction cost); $C(p)$ is the construction cost, H are the direct material and human losses in case of failure; γ_c is the discount rate; and $p_f(p)$ the structure's probability of failure. The design value p will be the value that maximises equation (2), given that $Z(p) > 0$.

5.2. Basic model and input data

The following models and cost data were used: $C(p) = C_0 + C_1 p\psi$, with $C_0 = \varphi\, 2 \times 10^6$; where $\psi = 1.25$ is a factor that accounts for the non-linear increase of cost as additional seismic resistance is provided to the structure;[4] φ is a factor that relates construction costs with the level of development (Table 2). The benefit was computed as $b = \alpha C_0$ (see α in Table 2). Values of parameter α are a function of the financial and technical risks, which in turn are closely related to the level of development.[29] Direct losses are divided into human losses and other losses, $H = H_{SLSC} + H_M$. The term H_M corresponds to all direct costs of losses and H_{SLSC} is the social cost of saving lives. The direct cost of losses is computed as $H_M = \delta\kappa C_1$, where $\kappa = 0.01$ is an estimated overall cost in addition to structural damage cost after failure and δ is a factor that varies depending upon the country; that is from 1.1 in the most developed to 0.75 in the less developed countries.

According to Rackwitz et al.,[19] $H_{SLSC} = SLSC\, k\, N_{PE}$ where k relates changes in mortality to changes in failure rate and N_{PE} is the number of people exposed (i.e. potentially endangered). The reference size of the exposed population in this study is taken as $N_{PE} = 1000$. SLSC can be interpreted as a compensation cost, for example the amount of resources destined by the social system to support the victims in an adverse event or the amount that should be covered by insurance. If it is assumed that the life risk is uniformly

Parameter		Average development groups		
		Developed	Developing	Underdeveloped
		Norway, USA, Canada, Japan, Australia, UK, Germany, France and so on	Argentina, South Africa, Malaysia, Chile, Colombia, Turkey and so on	Guatemala, India, Nigeria, Bolivia, Nicaragua, Mozambique, Kenya and so on
Expected benefit (% of C_0)	α	0.092 [0.074, 0.12]	0.146 [0.121, 0.166]	0.181 [0.151, 0.197]
Corrected discount rate[32,33]	γ_c	0.038 [0.023, 0.054]	0.078 [0, 055, 0, 10]	0.122 [0, 087, 0, 155]
Economic growth[31]	GDP/capita	33 158 [47 800, 24 200]	10 029 [15 000, 6 400]	3 375 [4 900, 1 200]
Life expectancy[31]	e	78.31 [81.25, 76.46]	71.50 [77.02, 42.45]	65.30 [78.55, 40.90]
% time devoted to raise the salary[29,30]	w	0.14 [0.13, 0.20]	0.16 [0.15, 0.20]	0.17 [0.18, 0.15]
% of GDP available for risk control[29,30]	g	50%	60%	70%
Basic construction cost factor[9]	φ	1.25	1.15	1
Health and long life[30]	SLSC	7.754×10^5 [5.104×10^5, 1.127×10^6]	2.562×10^5 [1.622×10^5, 4.143×10^5]	8.849×10^4 [2.543×10^4, 1.409×10^5]
Human development index[26]	HDI	0.944 [0.965, 0.912]	0.790 [0.863, 0.653]	0.620 [0.765, 0.311]
Results				
Mean optimum Design acceleration	p_{opt}	2.66 [2.53, 2.80]	2.39 [2.29, 2.58]	2.25 [2.09, 2.37]
*COV optimum design acceleration	COV(p)	2.6%	3.6%	4.0%
% difference with respect to code-based design parameter		10%	19%	24%

* COV: coefficient of variation

Table 2. Summary of data and optimum seismic design parameter for different levels of development

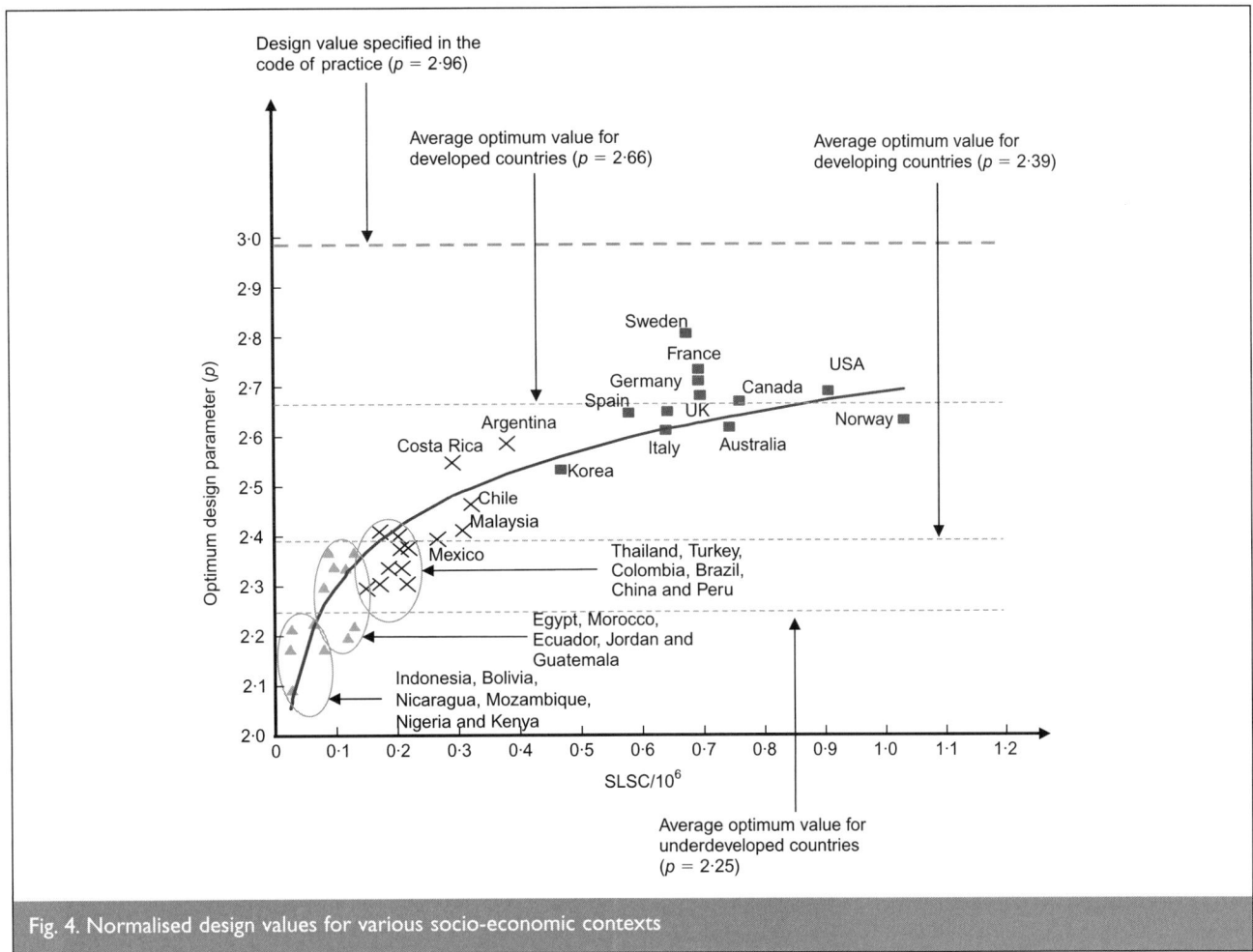

Fig. 4. Normalised design values for various socio-economic contexts

distributed over age and gender, the probability of actually being killed in case of the event, k, is estimated based on structural performance levels associated with life safety—that is, $k = 0.01$ (for further details see Rackwitz[30])

Regarding the economic assessment—that is, net present value evaluation—it is assumed that most of the project will be carried out by private investors and resources will be provided mainly by international multilateral institutions (e.g. World Bank). Therefore, γ_c is the 10-year corrected average interest rate, which attempts to capture the economic and political stability of the region. It corresponds to the discount rate affected by the cost of money when transactions are made in local currency. The values of γ_c are usually slightly larger than discount rates used in common practice.

Finally, in order to compute the probability of failure, it was assumed that both the resistance and the magnitude of disturbances are lognormally distributed. Although more elaborate models can be used, for the scope of the current paper this model was considered appropriate. Failure will occur if the demand exceeds the capacity; then, if V_R and V_S are the coefficients of variation of resistance and demand respectively, the failure probability can be expressed as

$$\boxed{3} \qquad p_f(p) = \Phi\left(-\frac{\ln\left[p\sqrt{(1+V_S^2)/(1+V_R^2)}\right]}{\sqrt{\ln\left[(1+V_S^2)(1+V_R^2)\right]}}\right)$$

where p describes the central safety factor (i.e. $p = \mu_R/\mu_S$) and, therefore, and indicator of the level of safety that must be provided. As countries become more developed, construction control and quality assurance become tighter. In this study, the variability in construction materials and processes was taken into account by assuming appropriate V_R values for every country; the average values for developed, developing and underdeveloped countries were 0.12, 0.175 and 0.25 respectively. The coefficient of variation of demand was assumed to be $V_S = 0.3$ for all cases.

5.3. Implication of results on sustainability

Figure 4 presents the relationship between the SLSC (as a measure of social development) and the socio-economic balanced (optimum) design criteria for the countries indicated. For the set of data considered, the SLSC and HDI are highly correlated ($\rho = 0.9$); similar results will therefore be obtained if the HDI is used instead of the SLSC. In addition, it was assumed that safety standards used by codes of practice in developing countries are mostly adaptations of those in developed countries. A reference central safety factor of $p = 2.96$ was therefore computed by assuming an overall reliability index of $\beta = 3.5$[5,34] with $V_S = 0.3$ and $V_R = 0.128$ (i.e. average values adopted for developed countries); the corresponding failure probability is $p_f = 2.326 \times 10^{-4}$.

The results in Fig. 4 show that there is a clear trend in which the optimum design parameter (i.e. central safety factor) increases as the level of social development increases. For less developed countries, small increments in the SLSC lead to significant differences in the optimum design parameter. By contrast, for developed countries this change is marginal. These results depend upon the initial construction costs, the expected benefits, the number of people exposed to the event, the choice of discount rate, and the parameters defining the probabilistic models. Nevertheless, the overall tendency presented in Fig. 4 will also be observed under similar conditions.

Comparing the average design criterion, p, for each level of development, it is observed that coefficients of variation are small and the differences vary between 10% and 24% (Table 2) with respect to the code specification (i.e. $p = 2.96$) assumed. These differences imply a reduction in the construction costs, which depend upon the structural material and the relationship between construction cost and resistance (i.e. $C(p) = C_0 + C_1 p^{\psi}$). Results show that less developed countries may save up to 20% of initial construction cost by taking a risk-balanced decision. If the design of housing uses average optimum values (i.e. $p = 2.25$ for underdeveloped and $p = 2.39$ for developing countries, see Fig. 4) rather than the values implied by international codes (i.e. $p = 2.96$), the savings in construction costs could be used, for instance, to increase the number of low-income houses constructed.

The criteria adopted for structural design and construction, together with the huge need for low-income housing and infrastructure in less developed countries, make the selection of appropriate safety standards a key component of sustainable development. By adopting safety standards specified in codes of practice of developed countries, less developed countries may be spending more to achieve levels of safety not warranted by their need for development. In other words, they may be expending more than what is an acceptable balance of risks in the context of the totality of all risks of living in those countries. Resources saved by rebalancing the risks can be directed towards building additional infrastructure or investing in other basic needs with safety levels that are in accordance with their socio-economic characteristics. Achieving sustainable development requires taking risks and balancing (optimising) the use of limited resources.

6. CONCLUSIONS

Sustainable development requires a balance between protecting the physical environment and its resources, and using these resources in a way that will allow the earth to continue supporting an acceptable quality of life. The construction industry plays a significant role in achieving sustainable development since it influences socio-economic development, in particular, in less developed societies. Engineers should focus on developing models able to use and process larger amounts of information obtained from different sources to provide solutions that have a positive impact on sustainability by encompassing the interests of larger sectors of society. By adopting safety standards specified in codes of practice in developed countries, less developed countries are spending more than what can be economically justified based on competing objectives of economic development and safety. Resources saved by understanding the context can be directed towards building more infrastructure or towards investing in other basic needs with safety levels that are in accordance with the local socio-economic status and objectives for economic development. Achieving sustainable development includes appropriate risk management and balancing (optimising) the use of limited resources.

REFERENCES

1. UN-CHIEF EXECUTIVES BOARD. Ensuring environmental sustainability. *One United Nations.* Ch. 2, 2005. See http://www.unsystemceb.org/oneun/2/p211.

2. WORLD BANK. *The Building Blocks of Development.* World Bank, New York, 2007. See Infrastructure webpage: http://web.worldbank.org/WBSITE/EXTERNAL/EXTABOUTUS/ORGANIZATION/EXTINFNETWORK/0,,menuPK:489896~pagePK:64158571~ piPK:64158630~theSitePK:489890,00.html

3. RACKWITZ R. Optimization and risk acceptability based on the life quality index. *Structural Safety*, 2002, **24**, No. 2–4, 297–331.

4. WEN Y. K. Reliability and performance based design. *Structural Safety*, 2001, **23**, No. 4, 407–428.

5. SÁNCHEZ-SILVA M. and ARROYO O. Comparing target spectral design acceleration values by using different acceptability criteria. *Structural Safety*, 2005, **27**, No. 2, 73–91.

6. HALL J., CRUICKSHANK I. and GODFREY P. S. Software supported risk management for the construction industry. *Proceedings of the Institution of Civil Engineers, Civil Engineering*, 2001, **144**, No. 1, 42–48.

7. BRUNDTLAND COMMISSION. *Our Common Future.* Oxford University Press, Oxford, 1987.

8. INTERNATIONAL COUNCIL FOR RESEARCH AND INNOVATION IN BUILDING AND CONSTRUCTION/UNITED NATIONS ENVIRONMENT PROGRAMME INTERNATIONAL ENVIRONMENTAL TECHNOLOGY CENTRE. *Agenda 21 for Sustainable Construction in Developing Countries.* CSIR Building and Construction Technology, Pretoria, South Africa, 2002.

9. UNITED NATIONS CENTRE FOR HUMAN SETTLEMENTS. *Promoting Sustainable Construction Industry Activities: First Consultation on the Construction Industry* (*Background Paper*), Tunisia UNCHS (HABITAT) and United Nations Industrial Development Organisation (UNIDO), Tunis, 1993.

10. UN-HABITAT. *Global Urban Observatory (GUO) and Human Settlements Statistics.* UN-Habitat, 2002. See: http://www.unhabitat.org/programmes/guo/table13.asp?countrycode.

11. PRASAD D. and HALL M. The construction challenge: sustainability in developing countries. *RICS Leading Edge Series.* 2004. Available from http://www.rics.org/NR/rdonlyres/53B43F2F-D7A5-4CA2-9144-8D5F5B928BAE/0/construction_challenge.pdf.

12. WORLD BUSINESS COUNCIL FOR SUSTAINABLE DEVELOPMENT. *The Cement Sustainability Initiative.* WBCSD, 2007. See http://www.wbcsdcement.org/concrete_misc.asp.

13. WORLD RESOURCES INSTITUTE. *World Resources 2000–2001.* WRI, 2001. See: http://www.wri.org.

14. INTERNATIONAL LABOUR ORGANISATION. *The Construction Industry in the Twenty-First Century: Its Image, Employment Prospects and Skills Requirements.* ILO, Geneva, 2001.

15. CONFEDERATION OF INTERNATIONAL CONTRACTORS' ASSOCIATIONS/UNITED NATIONS ENVIRONMENT PROGRAMME (CIC). *Industry as a Partner for Sustainable Development: Construction.* CICA and UNEP, UK, 2002.

16. WORLD BANK. See: http://web.worldbank.org/WBSITE/

EXTERNAL/NEWS/0,,contentMDK:20127296~
menuPK:34480~pagePK:34370~theSitePK:4607,00.html.

17. FRANGOPOL D. M., KALLEN M.-J. and NOORTWIJK M.
Probabilistic models for life-cycle performance of
deteriorating structures: review and future directions. *Steel
construction Progress in Structural Engineering Materials*,
2004, **6**, No. 4, 197–212.

18. SANTANDER C. F. and SÁNCHEZ-SILVA M. Design and
maintenance-program optimization for large infrastructure
systems. *Structure and Infrastructure Engineering:
Maintenance, Management, Life-Cycle Design &
Performance*. 2007, DOI: 10.1080/15732470600819104.

19. RACKWITZ R., LENTZ A. and FABER M. H. Socio-economically
sustainable civil engineering infrastructures by
optimization. *Structural Safety*, 2005, **27**, 187–229.

20. WEN Y. K. and KANG Y. J. Minimum building life-cycle cost
design criteria, I: methodology. *ASCE Journal of Structural
Engineering*, 2001, **127**, No. 3, 330–337.

21. LEVY H. *Stochastic Dominance: Investment Decision Making
under Uncertainty*. Kluwer Academic, Boston, 1988.

22. GODA K. and HONG H. P. Optimal seismic design
considering risk attitude, societal tolerable risk level and
life quality criterion. *ASCE Journal of Structural
Engineering*, 2006, **132**, No. 12, 2027–2035.

23. UNITED NATIONS CONFERENCE ON ENVIRONMENT AND
DEVELOPMENT. UNCED Report, 1992, A/CONF.151/5/Rev.1 13.

24. UNITED NATIONS DEPARTMENT OF SOCIAL AND ECONOMIC
AFFAIRS, DIVISION FOR SUSTAINABLE DEVELOPMENT. *Indicators
of Sustainable Development: Guidelines and Methodology*,
3rd edn. UN-DSEA, New York, 2007.

25. INTERNATIONAL SUSTAINABLE INDICATOR NETWORK. 2008. See
main page at: http://www.sustainabilityindicators.org/

26. UNITED NATIONS DEVELOPMENT PROGRAM. *Human
Development Report 2007/2008: Fighting Climate Change:
Human Solidarity In A Divided World*. UNDP, 2007. See
http://hdr.undp.org/.

27. NATHWANI J. S., LIND N. C. and PANDEY M. D. *Affordable
Safety by Choice: The Life Quality Method*. Institute for
Risk Research, University of Waterloo, Waterloo, Ontario
Canada, 1997.

28. CARMIGNANI F., COLOMBO E. and TIRELLI P. Consistency
versus credibility: how do countries choose their exchange
rate regime? *International Finance*, 2005, 0502001,
EconWPA.

29. SÁNCHEZ-SILVA M. and RACKWITZ R. Implications of the
high quality index in the design of optimum structures to
withstand earthquakes. *Journal of Structures, ASCE*, 2004,
130, No. 6, 969–977.

30. RACKWITZ R. The effect of discounting, different mortality
reduction schemes and predictive cohort life tables on risk
acceptability criteria. *Reliability Engineering and Systems
Safety*, 2006, **91**, No. 4, 469–484.

31. CENTRAL INTELLIGENCE AGENCY. *The World Fact Book*. CIA,
2007. See https://www.cia.gov/library/publications/
the-world-factbook/index.html.

32. DEUTSCHE BANK. Deutsche Bank research homepage, 2007.
See http://www.dbresearch.de/servlet/
reweb2.ReWEB?rwsite=DBR_INTERNET_EN-
PROD&$rwframe=0.

33. WORLD BANK. Data and Statistics webpage. See http://
www.worldbank.org/data/countrydata/countrydata.html.

34. APPLIED TECHNOLOGY COUNCIL. *ATC-34: A Critical Review of
Current Approaches to Earthquake-Resistant Design*. ATC,
California, 1995.

What do you think?

To comment on this paper, please email up to 500 words to the editor at journals@ice.org.uk

Proceedings journals rely entirely on contributions sent in by civil engineers and related professionals, academics and students. Papers should be 2000–5000 words long, with adequate illustrations and references. Please visit www.thomastelford.com/journals for author guidelines and further details.

Proceedings of the Institution of
Civil Engineers
Structures & Buildings 161
August 2008 Issue SB4
Pages 199–207
doi: 10.1680/stbu.2008.161.4.199

Paper 700054
Received 17/09/2007
Accepted 11/04/2008

Keywords: buildings, structures &
design/risk & probability analysis

Bruce R. Ellingwood
Raymond Allen Jones Chair
in Civil Engineering, Georgia
Institute of Technology,
Atlanta, USA

Structural reliability and performance-based engineering

B. R. Ellingwood PhD, PE, FASCE, NAE

Civil infrastructure facilities must be designed to
withstand demands imposed by their service
requirements and by natural environmental events. In
the past four decades, structural reliability methods have
matured to the point that they now provide the
framework for addressing safety and serviceability issues
in modern codified structural design, and most countries
have adopted such methods for code development.
While the normal design process usually results in a
constructed facility with a degree of integrity that is also
available to withstand challenges from unforeseen
events, events outside the traditional design envelope
may result in severe damage or economic losses and, in
extreme instances, precipitate catastrophic collapse. In
an era of heightened public awareness of infrastructure
performance, performance-based engineering (PBE)
offers a new paradigm that may allow structural
engineers to meet these challenges and to better match
building design with owner, occupant and social
performance expectations. Successful implementation of
PBE will require understanding and acceptance of the
supporting quantitative structural reliability risk
assessment methods and risk-informed decision tools
summarised in this paper.

1. INTRODUCTION

The design of civil infrastructure facilities to withstand
demands imposed by their service requirements and by natural
environmental events is guided by codes, standards and other
regulatory documents. In contrast to aerospace, automotive and
many marine structures, civil infrastructure facilities tend to be
one-of-a-kind, their performance may affect the lives of
thousands, the provisions governing their design are essentially
prescriptive and their design and construction are regulated by
public law. Although codes and standards provide the
minimum requirements for the structural design of facilities,
the normal design process usually results in a constructed
facility with a degree of integrity that is also available to
withstand challenges from unforeseen events. Events outside
the traditional design envelope, including extreme storms or
earthquakes, fires, accidents and malevolent attack, however,
may lead to severe damage or precipitate a catastrophic
collapse with extraordinarily severe consequences. Modern
design and construction practices and social forces have made
certain buildings and other infrastructure vulnerable to such
events. Public awareness of infrastructure performance and

safety issues has increased markedly during the past 30 years
as a result of media coverage of natural and man-made
disasters. The prospect of improving building practices to
achieve levels of facility performance beyond those provided
by minimum code requirements, to enhance facility robustness
and to lessen the likelihood of unacceptable damage from low-
probability, high-consequence threats now is receiving
heightened interest among engineers and other design
professionals. For example, a recent issue of the *Journal of
Performance of Constructed Facilities*, ASCE, 2006, Vol. 20, No.
4, was devoted to mitigating the potential for progressive (or
disproportionate) structural collapse.

Structural loads, material strengths, system strength and
stiffness and other factors in the building process are uncertain
in nature. Such uncertainties give rise to risk, which code-
writers must manage in the public interest. The safety level
inherent in any code, embedded in the various safety factors
chosen by the code developers, represents a value judgement
on acceptable risk based on past experience. This judgemental
approach to risk management worked reasonably well in an era
in which the technology in design and construction evolved
slowly. The rapidity of technological and social changes in the
21st century promises, however, to render this judgemental
approach obsolete.

The new paradigm of performance-based engineering (PBE) is
evolving to enable new building technologies and structural
design to better meet heightened public expectations and to
enable more reliable prediction and control of infrastructure
performance. This move must be supported by advances in
quantitative reliability and risk assessment tools. Structural
reliability theory has provided a quantitative link between the
practice of structural engineering and its social consequences,[1]
and promises to become even more important as a decision
support tool in PBE. This paper summarises some of the
advances made possible by structural reliability methods in
first-generation probability-based code development and
identifies new challenges that must be met for the profession to
advance, with PBE, to the next plateau.

2. PROBABILITY-BASED CODIFIED DESIGN

2.1. A brief review

Statistical tools have been used extensively since at least the
1930s to identify extreme environmental effects, such as wind

speed and precipitation, for design purposes.[2] The coupling of the demand or hazard side of the design equation with the response or capacity side through formal structural reliability analysis to achieve a quantitative metric of structural performance has occurred more recently.

The conceptual basis for structural reliability theory is relatively simple. In a departure from traditional deterministic thinking, the structural resistance or capacity, R, and structural actions owing to the applied loads, S, are modelled by random variables. Failure, defined by the engineer for the problem at hand, is defined by the probability that R is less than S, written as

$$\boxed{1 \qquad P_F = P[R < S] = \int [1 - F_S(x)] f_R(x) dx}$$

in which P_F is probability of failure to meet performance objectives identified by code, the designer or owner, $F_S(x) = P[S \leq x]$ is the cumulative distribution function of S and $f_R(x)$ is the density function of R. The term $1 - F_S(x) = P[S > x]$ represents the hazard, and will be denoted $H(x)$ in the sequel.

Until the late 1960s, developments in structural reliability were mainly theoretical in nature.[3] Beginning in the late 1960s and continuing through the end of the 1970s, however, the reliability and the design/standard communities began to move in step,[4] as research was undertaken to develop the first generation of practical probability-based limit states design (PBLSD) specifications. Modern probabilistic load modelling and first-order reliability methods date from that time, when most of the tools necessary to develop first-generation probability-based codes became available. Developments in the field of probability-based codified design have remained relatively stable in the past decade.

Probability-based limit states design codes worldwide all have adopted the general partial factors format

$$\boxed{2 \qquad \varphi R_n > \Sigma \gamma_i Q_{ni}}$$

in which R_n is the specified nominal (characteristic) strength, φ is the resistance factor, Q_{ni} represents nominal (characteristic) loads and γ_i is load factors. The product φR_n is denoted the design strength. The required strength, $\Sigma \gamma_i Q_{ni}$, is determined from the set of load combinations stipulated by the applicable code (e.g. ASCE Standard 7-05[4] or Eurocode No. 1[5]). The design format is transparently deterministic, but the load and resistance factors are based on explicit reliability benchmarks (reliability indices) obtained through a complex process of code calibration.[6] The first probability-based specification for steel structures in the United States (denoted load and resistance factor design, or LRFD) was introduced in 1986, and since has been followed by several other specifications. Elsewhere, code evolution in Canada, Mexico, Western Europe and Japan has proceeded concurrently and along similar lines.

2.2. Probability-based design: a critique
The move towards PBLSD entailed significant advances on several fronts. Perhaps most important, it required a thorough re-examination of the philosophical underpinnings of structural design, and led to a number of changes in the manner in which structural response is calculated for different limit states, particularly with regard to member and system stability. Technical bases for the provisions now are traceable, uncertainties are dealt with rationally and specific provisions can be (indeed, have been) revised in a systematic and scientific manner.

On the other hand, a number of important issues have yet to be addressed. For example, the target reliabilities were determined from first-order reliability analysis of a very large number of members and connections designed using existing traditional design criteria.[6] As structural system behaviour is rarely governed by individual member behaviour, these member reliabilities are no more than surrogates for the reliability of the structural system which, of course, is conditional on the information available and generally is unknown. System effects occasionally are hidden in the member safety checks; effective length factors (K) used in analysing stability of low-rise frames and response modification coefficients (R and C_d) that account for inelastic behaviour and energy dissipation in earthquake-resistant design are two examples. First-generation codes make little meaningful distinction between critical and non-critical structural elements. The basic requirements are still prescriptive, notwithstanding the probabilistic basis for the partial safety factors, and the fact that the partial factors are stipulated as constant to facilitate code implementation has led to a situation where uniform risk to populations across broad geographic areas cannot be assured, particularly for environmental and geophysical hazards. For example, the ASCE Standard 7-05 load combinations stipulate a (principal) factored wind load of $1.6 W_n$ for ultimate limit states design. The nominal wind speed used to calculate W_n for ordinary buildings in extra-tropical regions of the US historically has been based on a 50 year return period (3 s gust) wind velocity. The climatology in hurricane-prone coastal areas is vastly different from that in extra-tropical regions. Thus, in an attempt to achieve something close to uniform risk for buildings exposed to these different climates, the mapped wind speed in hurricane-prone regions in recent editions of ASCE 7-05 has been set equal to a 500 year return period wind speed (presumably representing an incipient failure event) divided by $\sqrt{1.6}$, a wind speed that has a return period substantially less than 50 years at some coastal sites. Engineers who are attempting to apply the wind load provisions intelligently find such manipulations difficult to understand. Moreover, the question of why the incipient failure wind event should be associated with a probability of 1 in 500 years or why it should be 1 in 500 year[8] for wind and 1 in 2475 years for earthquake[4] has yet to be addressed.

Finally, the structural reliabilities inherent to the first-generation PBLSD codes are related to social expectations of structural performance only insofar as the reliability benchmarks obtained from calibration to traditional practice can be validated. Dealing with novel structural systems and innovative construction methods for which there is little or no experience or designing for low-probability, high-consequence events outside the traditional design envelope is problematic when such practices and events are not reflected in the traditions on which the reliability benchmarks are founded.

3. PERFORMANCE-BASED ENGINEERING

First-generation PBLSD criteria are prescriptive, quantitative and detailed. Prescriptive criteria are unambiguous for the designer, are easily verified and enforced by a regulator, and are consistent with engineering pedagogy. On the other hand, prescriptive codes create the illusion that it is sufficient for the designer simply to meet the minimums of the code. In some cases, this can lead to a false impression of structural robustness; in others, a building in which the needs of the building occupants are not met. For example, many buildings and other structures designed by traditional prescriptive criteria failed to perform according to owner/occupant expectations during recent natural disasters, such as Hurricanes Hugo (1989), Iniki and Andrew (1992) and the Loma Prieta (1989) and Northridge (1994) earthquakes. Although the loss of life from these events was relatively low, suggesting that codes, by and large, are fulfilling their goal of assuring life safety, the direct economic losses from these five natural disasters exceeded $50 billion; the indirect losses and social disruptions are incalculable. At a time when public awareness of building and other infrastructure performance issues is increasing as a result of media coverage of such disasters, a better approach is required.

Performance-based engineering provides a new paradigm for structural design that is better able to match the completed building project with performance expectations of the building owner, occupants and public. PBE advances the philosophy of civil infrastructure design beyond the essential goal of public safety, providing engineers with more flexibility in designing with non-traditional systems and materials and in achieving innovative design solutions. The concept of PBE is not new, and indeed PBE is the *sine qua non* of structural design in the aerospace, automotive and marine industries. Its genesis in civil construction was in the late 1960s, when the US Department of Housing and Urban Development sponsored a large research programme at the National Bureau of Standards to develop criteria for design and evaluation of innovative housing systems. The recent resurgence of interest in PBE stems from a desire to tie structural design criteria more closely to the performance expectations of building owners and occupants, to take advantage of new building technologies, and to add value to the building process through innovation. Notable recent advances in PBE have occurred in two areas: fire engineering and earthquake engineering. In both areas, the motivating factors are strongly economic.

Fire protection traditionally has relied on component qualification testing according to ASTM Standard E119[7] or ISO Standard 834,[8] with acceptance criteria keyed to survival of a 'standard' fire for a required rating period. The qualification testing procedure is unambiguous and is used to determine design fire ratings for key structural components, expressed in terms of duration (typically 1–4 h) of a standard fire. On the downside, many of these test procedures and criteria have been in existence for nearly a century. They stipulate an unrealistic fire, address its impact on a component or compartment rather than a structural system and may penalise structural components and systems that are known to perform satisfactorily under realistic fire exposures.[9, 10] There are numerous initiatives underway on the international scene by organisations such as the Society of Fire Protection Engineers,

the National Fire Protection Administration, the American Society of Civil Engineers, the Council for Research and Innovation in Building and Construction, the European Convention for Constructional Steelwork aimed at advancing PBE concepts in fire-resistant structural engineering and design.

The recent advances in performance-based earthquake engineering (PBEE) are typified by Federal Emergency Management Agency (FEMA) 356 (2000)/American Society of Civil Engineers (ASCE) Standard 41-06,[11] the SAC project on steel moment frames (Structural Engineers Association of California, Applied Technology Council and California Consortium of Universities for Earthquake Engineering),[12] and the Structural Engineers Association of California (SEAOC) Vision 2000 effort.[13] Here, the motivation is three-fold: to limit the economic and social consequences of structural damage to owner/occupants and communities following an earthquake; to enhance building performance for clients who insist on a higher level of performance than can be guaranteed by current code minimums; and to better upgrade existing structures that are judged to be unsafe following an earthquake. Current research products and professional development activities have conveyed a measure of legitimacy to PBEE,[14] and in the past few years the approach has been followed in designing the seismic force-resisting systems of a number of tall buildings in the western US.

One common feature of most recent proposals for PBE is their distinction in levels of performance for different building categories where life safety or economic consequences differ. Current codes generally make such distinctions using an 'importance factor' (e.g. ASCE Standard 7-05), which simply leads to a higher design load, a step that may not lead to better performance and indeed may be irrelevant for dealing with certain low-probability events where effective design requires additional ductility rather than strength. In contrast, the design objectives in PBE are often displayed in a matrix (the International Code Coundil (ICC) Performance code[15] is an example), in which one axis describes severity of hazard (e.g. minor, moderate, severe) and the second axis identifies performance objectives (continued occupancy, life safety, collapse prevention). Facilities in categories where life safety or economic consequences differ are placed in appropriate bins in this matrix. The concept is illustrated in Fig. 1, using the categories that are defined in ASCE Standard 7-05 (I—low hazard to human life; II—normal; III—substantial hazard to human life (e.g. assembly facilities, schools); IV—essential (e.g. hospitals, police and fire stations, designated emergency shelters)). PBE might require that a hospital remain functional under an extremely rare event (sustaining minor damage), and provide continued service without interruption under a rare event. The shaded areas in Fig. 1 represent unacceptable performance. Current prescriptive codes essentially limit their focus to life safety under rare events. The approach represented by Fig. 1 is a more mature method for managing risk than using an importance factor, one that of course requires careful communication and mutual understanding among members of the design team rather than a simple reliance on prescriptive code provisions. The probabilities associated with event magnitudes in Fig. 1 are presented for illustration. They are consistent, in order of magnitude, with a

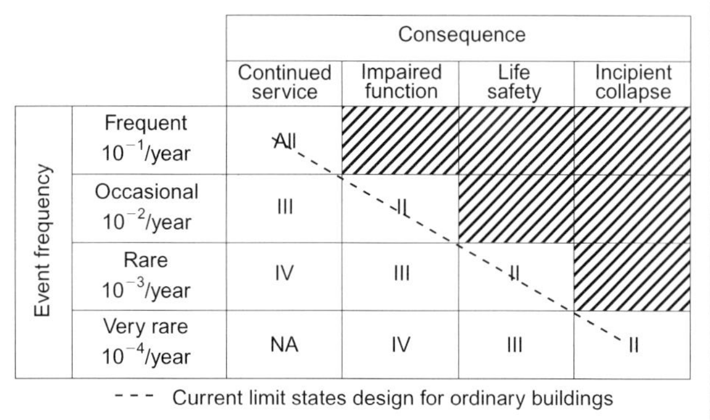

Fig. 1. Performance objectives and event probabilities for ASCE Standard 7-05 building categories[4]

number of PBE proposals, but have not been endorsed by any general code or standard group in the US. US Department of Energy (DOE) Standard 1020-02,[16] which governs design of certain critical facilities for the US Department of Energy, is an exception.

Risk-informed PBE is likely to be adopted (at least initially) for a relatively small subset of infrastructure projects where the nature of the hazard(s) or potential benefit to the project warrants the additional investment. Such decisions are likely to be based on project size, the nature of its anticipated use or occupancy, its prominence and the potential impact of a catastrophic failure on the surrounding community. Successful implementation of PBE in these circumstances will require careful articulation of performance objectives, establishment of relations between qualitatively stated performance goals and quantitative models of structural component and system behaviour and the development and validation of useable computational platforms, supporting databases and reliability-based tools to manage uncertainty and measure acceptable performance and risk.

4. FRAMEWORK FOR RISK-INFORMED ENGINEERING DECISIONS

Modern risk assessment of civil infrastructure has three general elements: hazard, consequences and context.[17] The hazard is a threat or peril—earthquake, fire, terrorist attack—that has the potential to cause harm. In some instances, the hazardous event (or spectrum of such events) can be defined in terms of annual frequency. More often than not, however, the mean annual frequency of the threat may be unknown or it may not be amenable to traditional statistical modelling techniques, and it is necessary to envisage a set of hazardous scenarios, without regard to their probability or mean annual frequency of occurrence.[18] The occurrence of the hazard has consequences—damage to or collapse of the constructed facility, personal injury, direct and indirect economic losses, damage to the environment—which must be measured by an appropriate metric reflective of the decision maker's value system. Finally, there is the context: individuals or groups at risk and decision makers concerned with managing risk may have different value systems and adopt different positions on how investments in risk reduction must be balanced against available resources.

4.1. Quantitative measurement of risk

Quantitative measures of the diverse risks to civil infrastructure are required to distinguish risks that are significant from those that are trivial and to achieve ordinal rankings of decision preferences.[19,20] The basic mathematical framework for risk assessment of a constructed facility is provided by the familiar theorem of total probability

$$3 \quad \begin{array}{c} P[\text{Loss} > \vartheta] = \Sigma_H \Sigma_{LS} \Sigma_{DS} \\ P[\text{Loss} > \vartheta | DS] \, P[DS|LS] \, P[LS|H] \, P[H] \end{array}$$

The term $P[H]$ is annual probability of occurrence of hazard H; $P[LS|H]$ is conditional probability of a structural limit state (yielding, fracture, instability), given the occurrence of H; $P[DS|LS]$ is conditional probability of damage state DS (e.g. negligible, minor, moderate, major, severe) arising from structural damage, and $P[\text{Loss} > \vartheta | DS]$ is probability of loss exceeding ϑ, given a particular damage state; and $P[\text{Loss} > \vartheta]$ is annual probability of loss exceeding ϑ. (For rare events, $P[H]$ and $P[\text{Loss} > \vartheta]$ are numerically equivalent to the annual mean rates of occurrence of these events, λ_H and $\lambda_{\text{Loss} > \vartheta}$, which are more easily estimated from the data maintained by public agencies.) If the hazard is defined in terms of a scenario (or set of scenarios), the risk assessment equation becomes

$$4 \quad \begin{array}{c} P[\text{Loss} > \vartheta | \text{Scenario}] = \Sigma_{LS} \Sigma_{DS} \\ P[\text{Loss} > \vartheta | DS] \, P[DS|LS] \, P[LS|\text{Scenario}] \end{array}$$

The parameter ϑ is a loss metric: number of injuries or death, damage costs exceeding a fraction of overall replacement costs, loss of opportunity costs and so on, depending on the objectives of the assessment. Underlying equations (3) and (4) is the assumption that the damage state probability is dependent on the limit state but not directly on the hazard; for example the probability that damage is 'moderate' may depend on the limit state 'yielding', but not on the hazard that caused yielding.

At the heart of risk-informed PBE is identification of dominant contributors to risk.[21,22] The risk of low-probability events with trivial consequences is negligible, while the risk of events with similar likelihood but with grave consequences may be unacceptable. Equations (3) and (4) deconstruct the risk assessment into its major constituents and along disciplinary lines, making the process of risk assessment and management transparent. Engineering strategies seldom have an impact on λ_H, which is affected by other means such as changing the facility site or controlling access to it, installing protective barriers, proscribing hazardous substances and so on. An analysis of the frequencies of competing hazards allows trivial hazards to be screened from the decision process and appropriate risk mitigation strategies to be devised for those hazards that lead to unacceptable increases in building failure rates above the *de minimis* level, the threshold below which societies normally do not impose any regulatory requirements for risk management. This level is thought to be on the order of 10^{-7} to 10^{-6}/year.[23] This screening process is illustrated in Fig. 2 for two generic hazards. Hazard H_1 may be significant at low intensities (governing continued function) but inconsequential for high intensities (governing life safety or incipient collapse), at which

H_2 becomes dominant. Engineering design directly impacts the probabilities $P[LS|H]$, $P[DS|LS]$ and $P[LS|Scenario]$. Finally, the conditional probability, $P[Loss > \vartheta|DS]$, is best determined by the building owner (or manager) and insurance underwriter, as it involves costs of repairing the damaged structure, estimation of losses in revenue or business opportunity and the cost required to insure those losses. If the scenario approach (equation (4)) rather than the fully coupled assessment (equation (3)) is adopted because of a lack of data on hazard frequency, the loss (risk) estimate is conditional. Competing risks from postulated scenarios may be evaluated, but the losses cannot be benchmarked meaningfully against competing risks associated with other hazards.

Cost-effective PBE design requires appropriate attention to all terms in equations (3) and (4). All sources of uncertainty, from the hazard occurrence frequency to the response of the system, must be considered, propagated through the risk analysis framework defined by equations (3) and (4), and displayed clearly to obtain an accurate picture of risk.

4.2. Practical risk-based performance assessment

Risk-informed PBE for explicit levels of performance requires a connection between probability (frequency) of hazard occurrence, structural limit state, and expected performance level. The necessary ingredients for the assessment process are probabilistic models of the spectrum of hazards and of the capacity of the system to withstand the demands resulting from those hazards. The level of analysis required depends on the performance requirements of the project. The analysis need not be especially complex, as shown in the simple illustration that follows.

As defined in equation (1), the spectrum of hazards can be expressed through a complementary probability distribution, $H(x) = P[X > x]$, describing the likelihood that specific levels of demand are exceeded. It has been found that for many hazards with return periods of interest in civil infrastructure performance assessment (on the order of 100–10 000 year), the hazard often can be expressed approximately as

$$5 \qquad \ln H(x) \approx \ln \alpha - k \ln x$$

in which α, k are parameters.

The capacity of a system to withstand demands from such a

hazard is modelled by a limit state probability, or 'fragility,' in which the conditioning variable is the specific level of demand, x. The fragility often is described by a lognormal distribution[24-26]

$$6 \qquad P[LS|X = x] = \Phi[(\ln x - m_R)/\beta_R]$$

in which m_R is median capacity and β_R is logarithmic standard deviation of capacity. The family of fragility curves in Fig. 3, adapted from a seismic fragility analysis of a six-storey steel braced frame building in mid-America subjected to earthquake ground motion,[27] depict the damage limit state probabilities of the frame for progressively more severe limit states as a function of earthquake intensity (characterised by the spectral acceleration of the building at its fundamental period, S_a). The damage state probabilities are determined, if required, for vulnerability and loss assessment by relating the limit state (e.g. excessive inelastic deformation) to heuristically defined damage states. For example, the 'immediate occupancy' damage state probability in Fig. 3 was determined as the probability that the maximum interstorey drift angle in the frame exceeded 0·01, a point at which non-linear action initiates in the frame, accompanied by damage to non-structural partitions and other components. While developing these fragilities required a series of simulations involving non-linear dynamic analyses with ensembles of earthquake ground motion, such an analysis is well within the capabilities of engineers who are tasked with developing codes. Furthermore, such fragilities (more precisely, their parameters) can be compiled conveniently for different types of construction. An example of such a compilation can be found in chapter 5 of the Hazus Technical Manual,[28] Table 5·9, for 36 types of building structures, ranging from single-storey residential wood construction to 12-storey reinforced concrete or steel building frames, subjected to earthquakes of various intensities.

The limit state probability can be obtained by convolving equations (5) and (6) (see also equation (3) for context). To first approximation, this convolution leads to

$$7 \qquad \begin{aligned} P[LS] &= \Sigma_H \, P[LS|H] \, P[H] \\ &= H(m_R) \exp{[(k\beta_R)^2/2]} \end{aligned}$$

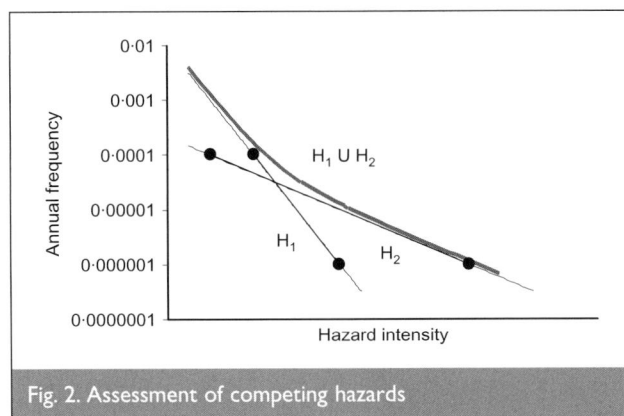

Fig. 2. Assessment of competing hazards

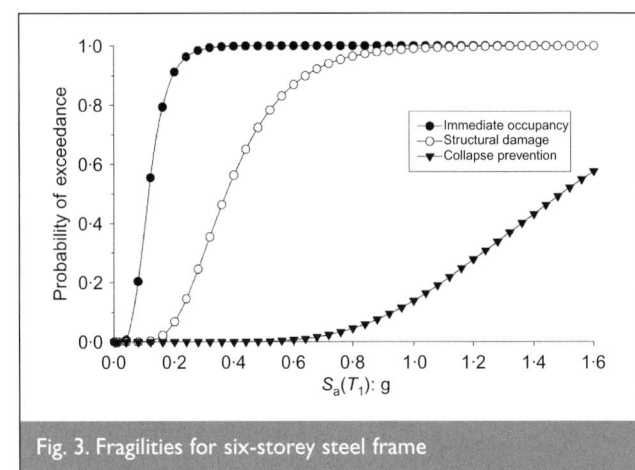

Fig. 3. Fragilities for six-storey steel frame

In words, equation (7) states that the limit state probability equals the hazard evaluated at the median capacity multiplied by a correction factor which is determined by the slope (uncertainty) in the hazard, k, and the slope (uncertainty) in the system fragility. Using equation (7), relatively simple risk-based design criteria can be derived that utilise hazard information as it currently is presented in codes and, at the same time, is consistent with stipulated performance requirements.

Current design for extreme environmental or geophysical hazards often is based on a 'uniform hazard,' provided in the form of a hazard map defining demand, S_d for specified $P[S > S_d] = H(S_d)$ [say, 2% probability of exceedence in 50 years or 2% per year]; maps of annual extreme wind speeds, snow loads, or earthquake accelerations will be familiar to many readers. Suppose that the design resistance, R_d, of the system is stipulated at the fifth percentile value of the fragility (see Fig. 3 and equation(6)). The mapped hazard can be used in the customary way to design for a targeted risk with the simple equation

$$8 \quad R_d = DF \times S_d$$

in which DF is the design factor. The lognormal fragility model implies that R_d is related to the median, m_R, by

$$9 \quad R_d = m_R \exp\left[-1{\cdot}645\,\beta_R\right]$$

Solving equation (9) for m_R, substituting into equation (7), and noting that the ratio of $H(S_d)/P[LS]$ is typically on the order of 10, a simple approximation to DF can be derived

$$10 \quad DF \approx 2{\cdot}6/\sqrt{k}$$

in which the error in neglecting the dependence of DF on β_R is approximately 7% for the typical range $0{\cdot}3 \leqslant \beta_R \leqslant 0{\cdot}6$. Choosing typical values of $\beta_R = 0{\cdot}40$ and $k = 4$ for a building exposed to seismic hazard (see, for example, Hazus Technical Manual[28]), DF = 1·3. In other words, assuming that the characteristic (nominal) design resistance is comparable to the 5-percentile value of strength or capacity, the facility should be designed for a demand corresponding to a (mapped) 10^{-3}/year hazard, multiplied by the factor 1·3 to achieve a risk less than 10^{-4}/year. A hazard of 10^{-3}/year corresponds to a return period of 1000 years. This return period is selected for illustrative purposes (see, for example, Fig. 1), but is typical as a design basis for extreme environmental hazards (e.g. DOE Standard 1020-02[16]).

This simple example highlights the difference between uniform hazard and uniform *risk* approaches to design. Recent proposals for PBE strive to make this distinction. A similar approach was followed in developing the fundamental design requirements in DOE Standard 1020-02[16] and ASCE Standard 43-05,[29] which are applicable to a much narrower range of facilities but takes a similar approach, identifying target performance goals in terms of decreasing annual probabilities for structures, systems and components in different design categories (related to severity of failure consequences) exposed to natural hazards such as wind, earthquake and flood.

5. RISK MEASUREMENT, TOLERANCE AND COMMUNICATION

For those projects adopting a risk-informed approach to PBE, one of the most difficult challenges will be to develop quantitative measures of risk and to communicate those quantitative risk benchmarks selected for the project to decision makers and regulators who may find such concepts difficult to grasp.

5.1. Risk measurement

The notion of measuring risk in structural design through probabilities (or reliability indices) or expected losses has become acceptable in the structural engineering and building code communities only relatively recently.[1,30] First-generation probability-based limit states design criteria for buildings and other structures, such as ASCE Standard 7-05,[4] are based on a notional benchmark limit state probabilities (related to 'reliability indices' in so-called first-order reliability analysis), which were determined through a complex process of calibration to existing engineering practice rather than *ab initio*. The reliability index, β, is a simple risk measurement and communication tool, particularly suited to situations in which uncertainties in the underlying probabilistic modelling process are large and relevant probabilities for decision making are small and difficult to interpret. It is unlikely that probability-based limit states criteria would have been accepted in the 1980s by professional structural engineers, let alone the building codes, had they been justified on the basis of a benchmark probability of 10^{-5}/year. The reliability index is more easily computed and understood and avoids the difficulty that is inherent to explaining low-probability events. More sophisticated risk measurements through expected losses—deaths and injuries, direct and indirect economic losses—have been conducted for certain very large or unique projects, where the consequences of failure are catastrophic, the cost of the project warrants the additional investment in the risk analysis, or the regulatory authority requires it. Risk-informed decision making also has a longer history in design and development of critical facilities such as nuclear power plants, chemical refineries and process facilities, liquid natural gas (LNG) storage facilities, offshore platforms, and similar facilities.

It is often important to convey some sense of the accuracy (or confidence) in the measurement. Knowledge-based (or epistemic) uncertainties in the assessment of performance of an infrastructure facility subjected to a rare event usually are substantial and should be reflected in any risk-informed decision because they determine the level of confidence that can be placed in the assessment. The impact of epistemic uncertainty on the risk analysis framework can be visualised as causing the probability models describing hazard, structural capacity, damage and loss to be, themselves, random, with relative frequencies that define the relative plausibility of the models. As a result, the risk estimate (damage probability, expected loss) is characterised by a frequency distribution, the dispersion of which displays the epistemic uncertainty in the estimate and conveys a sense of the accuracy in the risk assessment. If the epistemic uncertainties in the risk assessment are small, the frequency distribution is centred on the point estimate; conversely, if the epistemic uncertainties are large, the frequency distribution of $P[LS]$ will be broad. This concept is illustrated in Fig. 4, which displays the epistemic uncertainty associated with the assessment of the risk of structural collapse of the building illustrated in Fig. 3 exposed to earthquake hazard in the region near Memphis, TN in the US. The mean value is 1.1×10^{-4}/year, while the 95% interval is $0.3-2.7 \times 10^{-4}$/year (approximately one order of magnitude). The mean is an unambiguous point estimate of risk,[16,29] is a natural choice in performing minimum expected cost (or loss) analyses, and there are decision-theoretic justifications for selecting it. A statement of confidence in the results of the risk assessment is, however, often a useful risk communication tool.

5.2. Risk acceptance

Tolerance and acceptance of risk depend on the decision maker and context of the decision. Acceptable risk can be determined only in the context of what is acceptable in other activities, what investment is required to marginally reduce the risk, and what losses might be incurred if the risk were to increase. Most individuals are risk-averse (implying that they require a substantial benefit in return for accepting marginal increases in risk), while governments and large corporations tend to be risk-neutral.[31,32] Recent studies, summarised by Corotis,[33] have indicated that acceptance of risk is based more on the perception of risk than on the actual probability of occurrence and that any biases in perception shape decisions. While probability of occurrence is a significant component of risk, other factors also are important. For example, most societies view incidents involving large numbers people differently from incidents involving individuals. Reid[34] has suggested that individuals view risks as negligible if comparable to mortality risk from natural hazards (on the order 10^{-6}/year) and as unacceptable if comparable to mortality from disease (on the order 10^{-3}/year in the 30 to 40 age group). Although it is natural to use mortality statistics from disease and accidents to benchmark risk (e.g. automobile traffic fatalities in the United States have stabilised at approximately 2×10^{-4}/year for many years), comparisons of annual frequencies must

consider differences in exposure and consequences, and attempts to correct for such effects are subjective. The question of what constitutes acceptable risk in the built environment has not been answered definitively. Current codes appear to be delivering building products that are acceptable from a life-safety point of view. The threshold of acceptable risk depends on one's point of view; to the user of a facility, any risk below the (unknown) threshold is acceptable, while to the developer any risk above the threshold represents wasted cost. These different viewpoints must be reconciled as part of the PBE design process.

5.3. Risk communication

Risk-informed PBE requires a continuing dialogue among the members of the project team and other project stakeholders to understand basic issues and enhance the credibility and acceptance of the results of the risk assessment. A successful risk assessment can be scrutinised independently by a peer reviewer and provides an easily understood audit trail leading to key decisions affecting project safety and other performance objectives. The performance objectives and metrics must be clearly identified and agreed upon, and uncertainty analysis should be a central part of the decision model. To most decision makers, ranging from engineers and facility managers with professional training to elected officials representing the public, it is the estimated consequences (deaths or injuries, direct economic losses, and deferred opportunity losses) that are important. Trade-offs that occur between investment in PBE and risk reduction must be treated candidly, and the entire decision process must be made as transparent as possible.

6. CONCLUDING REMARKS

Structural reliability analysis has provided structural engineers and code writers with a framework for advancing engineering design practice through first-generation probability-based limit states design codes. Looking beyond the first decade of the 21st century, the worldwide move towards PBE and developments in computation support for structural design will be accompanied by a corresponding move towards improved quantitative risk and reliability bases for design. The structural engineering profession, which resisted the complex notions of probability and risk when first-generation probability-based codes were first implemented three decades ago, now appears willing to consider such concepts and bases for certain building projects. The structured thinking required to develop such bases will provide a framework for forecasting or assessing the performance of a structure threatened by natural and man-made hazards, facilitate a more balanced allocation of professional and financial resources for reducing or managing risks, and would enhance the ability of the profession to identify and respond to potential new challenges in structural design. To be successful, PBE requires an acknowledgement that proper engineering design involves looking beyond minimum code requirements. The project team must acknowledge that uncertainty in achieving the project performance goals and objectives cannot be eliminated entirely; that risks presented by events outside the customary design envelope cannot be avoided; but that reduction in risk can be achieved through both technical and non-technical measures by additional

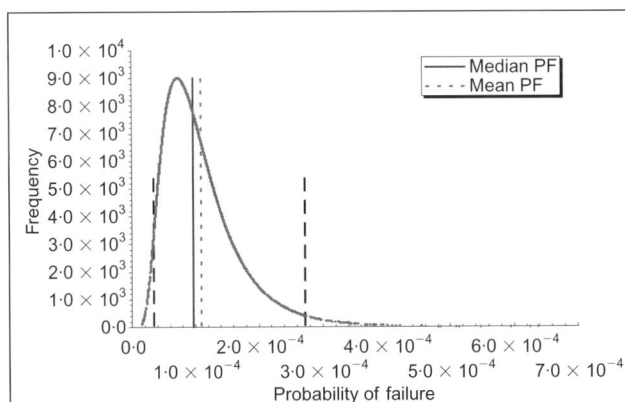

Fig. 4. Frequency distribution of damage state probability for braced frame in Fig. 3, showing the mean, median and 95% confidence interval on estimated damage state probability

investment in design and construction practices that enhance robustness and resilience of civil infrastructure systems.

ACKNOWLEDGEMENTS

The views expressed in this paper are based on research supported for many years by several organisations, including the US Nuclear Regulatory Commission, Oak Ridge National Laboratory, the US Army Corp of Engineers, the National Science Foundation, Sandia National Laboratory, the National Institute of Building Sciences, and the Mid-America Earthquake Center, an NSF-supported research center funded at the University of Illinois at Urbana-Champaign. This support is gratefully acknowledged. The views expressed are, however, solely those of the author, and may not represent the positions of the sponsoring organisations.

REFERENCES

1. ELLINGWOOD B. R. Probability-based codified design: past accomplishments and future challenges. *Structural Safety*, 1994, **13**, No. 3, 159–176.

2. DAVENPORT A. G. Rationale for determining design wind velocities. *Journal of Structural Division ASCE*, 1960, **86**, No. 5, 39–68.

3. FREUDENTHAL A. M., GARRELTS J. and SHINOZUKA M. The analysis of structural safety. *Journal of Structural Division ASCE*, 1966, **92**, No. 1, 267–325.

4. AMERICAN SOCIETY OF CIVIL ENGINEERS. *ASCE/SEI Standard 7-05 Minimum design loads for buildings and other Structures*. ASCE, Reston, VA, 2005.

5. COMITÉ EUROPÉEN DE NORMALISATION. *Eurocode 1: Actions on Structures* (EN 1991-1). CEN, Brussels, 2006.

6. ELLINGWOOD B. and GALAMBOS T. V. Probability-based criteria for structural design. *Structural Safety*, 1982, **1**, No. 1, 15-26.

7. AMERICAN SOCIETY FOR TESTING AND MATERIALS. *Standard Methods of Fire Tests of Building Construction and Materials (ASTM Standard E119-02)*. ASTM, Philadelphia, PA, 2002.

8. INTERNATIONAL ORGANIZATION FOR STANDARDIZATION. *Standard 834: Fire Resistance Tests – Elements of Building Construction*. ISO, Geneva, 1994.

9. BAILEY C. G., LENNON T. and MOORE C. B. The behaviour of full-scale steel-framed buildings subjected to compartment fires. *The Structural Engineer*, 1999, **77**, No. 8, 15–21.

10. WANG Y. C. An analysis of the global structural behavior of the Cardington steel-framed building during the two BRE fire tests. *Engineering Structures*, 2000, **22**, No. 5, 401–412.

11. FEDERAL EMERGENCY MANAGEMENT AGENCY. *NEHRP Guidelines for the Seismic Rehabilitation of Buildings (FEMA 356)*. FEMA, Washington, DC, 2000. (Recently published as *ASCE Standard 41-06*).

12. CORNELL C. A, JALAYER F., HAMBURGER R. O. and FOUTCH D. A. Probabilistic basis for 2000 SAC FEMA Steel Moment Frame Guidelines. *Journal of Structural Engineering ASCE*, 2002, **128**, No. 4, 526–533.

13. STRUCTURAL ENGINEERS ASSOCIATION OF CALIFORNIA. *Vision 2000: Performance based seismic engineering of buildings*. SEOC, Sacramento, CA, 1995. (Also published in *SEAOC Blue Book 1999*, Appendix G, Conceptual framework for performance-based seismic design.)

14. HAMBURGER R. O. Implementing performance-based seismic design in structural engineering Practice. *Proceedings of 11th World Conference on Earthquake Engineering*, 1996 (Paper No. 2121), Elsevier Science, Oxford.

15. INTERNATIONAL CODE COUNCIL. *ICC performance code for buildings and facilities*. ICC, Country Hills, IL, 2006.

16. US DEPARTMENT OF ENERGY. *DOE STD 1020-2002 - Natural phenomena hazards design and evaluation criteria for Department of Energy facilities*. US Department of Energy, Washington DC, 2002.

17. ELMS D. G. Risk assessment. *Engineering Safety* (D. BLOCKLEY (ed.)). McGraw-Hill International, Berkshire, UK, 1992, pp. 28–46.

18. GARRICK B. J., HALL J. E., KILGER M., MCDONALD J. C., O'TOOLE T., PROBST P. S., PARKER E. R., ROSENTHAL R., TRIVELPIECE A. W., VAN ARSDALE L. A. and ZEBROSKI E. L. Confronting the risks of terrorism: making the right decisions. *Reliability Engineering and System Safety*, 2004, **86**, 129–176.

19. STEWART M. G. and MELCHERS R. E. *Probabilistic Risk Assessment of Engineering Systems*. Chapman and Hall, London, UK, 1997.

20. VRIJLING J. K., VAN HENGEL W. and HOUBEN R. J. Acceptable risk as a basis for design. *Reliability Engineering and System Safety*, 1998, **59**, 141–150.

21. ELLINGWOOD B. R. and WEN Y. K. Risk-benefit based design decisions for low probability/high consequence earthquake events in Mid-America. *Progress in Structural Engineering and Materials*, 2005, **7**, No. 2, 56–70.

22. ELLINGWOOD B. R. Strategies for mitigating risk to buildings from abnormal load events. *International Journal of Risk Assessment and Management*, 2007, **7**, Nos 6/7, 828–845.

23. PATE-CORNELL E. Quantitative safety goals for risk management of industrial facilities. *Structural Safety*, 1994, **13**, No. 3, 145–157.

24. SINGHAL A. and KIREMIDJIAN A. S. Method for probabilistic evaluation of seismic structural damage. *Journal of Structural Engineering*, ASCE, 1996, **122**, No. 12, 1459–1467.

25. SHINOZUKA M., FENG M. Q., LEE J. and NAGANUMA T. Statistical analysis of fragility curves. *Journal of Engineering Mechanics* ASCE, 2000, **126**, No. 12, 1224–1231.

26. WEN Y. K. and ELLINGWOOD B. R. The role of fragility assessment in consequence-based engineering. *Earthquake Spectra EERI*, 2005, **21**, No. 3, 861–877.

27. KINALI K. and ELLINGWOOD B. R. Seismic fragility assessment of steel frames for consequence-based engineering: A case study for Memphis, TN. *Engineering Structures*, 2007, **29**, No. 6, 1115–1127.

28. FEDERAL EMERGENCY MANAGEMENT AGENCY. *Multi Hazard Loss Estimation Methodology: HAZUS Technical Manual*. FEMA, Washington, DC, 2003.

29. AMERICAN SOCIETY OF CIVIL ENGINEERS. *ASCE/SEI Standard 43-05 Seismic Design Criteria for Structures, Systems and Components in Nuclear Facilities*. ASCE, Reston, VA, 2005.

30. ELLINGWOOD B. R. Acceptable risk bases for design of structures. *Progress in Structural Engineering and Materials*, 2001, **3**, No. 2, 170–179.

31. SLOVIC P. *The Perception of Risk*. Earthscan Publications, Sterling, VA, 2000.
32. FABER M. and STEWART M. G. Risk assessment for civil engineering facilities: a critical appraisal. *Reliability Engineering and System Safety*, 2003, **80**, 173–184.
33. COROTIS R. B. Socially relevant structural safety, *Applications of Statistics and Probability in Civil Engineering, Proceedings of ICASP 9* (DER KIUREGHIAN A., MADANAT S. and PESTANA J. (eds)). Millpress, Rotterdam, 2003, pp. 15–26.
34. REID S. G. Acceptable risk criteria. *Progress in Structural Engineering and Materials*, 2000, **2**, No. 2, 254–262.

What do you think?

To comment on this paper, please email up to 500 words to the editor at journals@ice.org.uk

Proceedings journals rely entirely on contributions sent in by civil engineers and related professionals, academics and students. Papers should be 2000–5000 words long, with adequate illustrations and references. Please visit www.thomastelford.com/journals for author guidelines and further details.

Proceedings of the Institution of
Civil Engineers
Structures & Buildings 161
August 2008 Issue SB4
Pages 209–214
doi: 10.1680/stbu.2008.161.4.209

Paper 700043
Received 06/08/2007
Accepted 24/04/2008

Keywords: codes of practice &
standards/risk & probability analysis/
safety & hazards

Ton Vrouwenvelder
Professor, Delft Technical
University, The Netherlands

Treatment of risk and reliability in the Eurocodes

T. Vrouwenvelder

Designing a structure essentially comes down to making decisions in the face of uncertainty. This paper discusses how the new generation of European standards, the Eurocodes, deals with this issue. According to *EN 1990 Eurocode-Basis of Structural Design*, the theory of probability may be considered as the appropriate design tool in this respect. To keep the design process practicable, however, the probabilistic methods are translated into a verification process on the basis of partial factors. In setting values for the partial factors, theoretical considerations and long-term design experience are mixed. Next to the design for normal loading conditions, the Eurocode gives special attention to the design for accidental situations such as fire, explosions or errors. When dealing with these topics the code is of a more informal nature, giving freedom to designers and authorities. The recommendations range from prescribed solutions (minimal ductility and redundancy, horizontal and vertical tyings, etc.) to a formal risk analysis for important or extraordinary structures. Non-structural measures, extra safety for key elements and alternative load paths are presented as valuable elements in the safety strategy. The code reflects the increasing awareness of the engineering society for the importance of risk-based decision making in design and maintenance.

1. INTRODUCTION

The Eurocodes form a coherent and operational set of structural building standards for all common materials, including soil, and all types of civil engineering structures. The codes are performance based in the sense that a clear distinction is made between the requirements on the one hand and solutions on the other. The basic requirements, of course, are connected with safety, serviceability and economy for the anticipated lifetime of the structure.

As with most present-day codes, the Eurocodes are based on the limit state concept and on a set of characteristic values and partial factors. A limit state is the demarcation between desired and adverse states of the structure. Characteristic values for loads and resistances, where possible, correspond to prescribed probabilities and return periods. The partial factors take care of uncertainties and scatter and are tuned to desired levels of reliability.

This type of design is referred to as a probability-based design procedure. It should be considered as a consistent simplification of fully probabilistic methods.[1–3] In the current paper the link between the methods will be discussed. It should be noted that in *EN 1990 Eurocode-Basis of Structural Design*, the fully probabilistic method is mentioned as a useable alternative, but the relevant authorities are invited to give specific conditions. In annex C of EN 1990 only basic information is given.

Next to limit state design, the word 'robustness' is a key word in safety-related design. The notion of robustness is that a structure should not be too sensitive to local damage, whatever the source of damage: design errors, building errors, lack of maintenance, extreme loads, intentional damage and so on. Depending on the type of structure, various measures or levels of analysis are required. For low-class buildings there are no requirements. For the highest class of structures the designer is invited to carry out a formal qualitative or even quantitative risk analysis. More details will be discussed further in this paper.

2. RELIABILITY REQUIREMENTS

The first clause in EN 1990, section 2, requirements, states that

> A structure shall be designed, and executed in such a way that it will, during its intended life, with appropriate degrees of reliability and in an economical way sustain all actions and influences likely to occur during execution and use, and remain fit for the use for which it is required.

Given this requirement there is a need to distinguish between desired and undesired performance. The borderline between the two types of behaviour is called a limit state. Two main categories of limit states are distinguished.

(a) *Ultimate limit states* are associated with collapse, either on the level of structural members or, in particular, in relation to accidental limit states, on the level of a structural system.

(b) *Serviceability limit states* are associated with the usefulness of the structure. A further distinction is possible into reversible and irreversible serviceability limit states.[4] In the first case the limit state is no longer exceeded when the actions are removed, as for instance may happen for large elastic deflections and excessive vibrations. For irreversible limit states the exceedance will remain even when the actions are removed, as is for instance the case for most cracks.

The basic requirement for structural design is that 'limit states should not be exceeded with appropriate degrees of reliability'. This is an important notion in the performance-based design: the limit state requirements are not absolute, but a small probability of failure is accepted explicitly. In this way, doors are opened to a variety of solutions where designers can follow their own preferences as long as the basic reliability requirement is met. For example: structural fire safety may be obtained by sprinkler installations, by fireproof claddings, by strong members or a combination: the only point is that the failure probability should be below some target value. In a similar way, materials may show a small scatter (expensive quality control) or a large scatter (cheap production) as long as corresponding proper partial factors are used to meet the reliability requirement.

Numerical values for the acceptable failure probabilities are specified in annexes B and C of the code. It is stated, however, that probabilities should be considered as notional values that do not necessarily represent failure rates, but are used as operational values for code calibration purposes and comparison of reliability of structures. The strong point of probabilistic modelling and decision making is strong logic rather than accurate description of reality. Given a model, the set of design decisions can be rational and consistent. For the discrepancy between failure probabilities in theory and practice a number of reasons can be formulated. One is the systematic overdimensioning of structures for various reasons such as simplicity of design and conservatism in, or inadequacy of, models. Another one, of course, is the lack of data and the subsequently necessarily subjective estimation of statistical parameters. A third is the omission of human error and the inevitable unintended consequences of human decision making. For a more thorough discussion on these matters, see Ditlevsen[5,6] Elms and Turkstra.[7]

3. THE PARTIAL FACTOR METHOD

The safety and serviceability requirements of the Eurocodes are considered to be fulfilled by applying the partial factor method. The design formats and load combination rules for this method are shown in EN 1990. Representative values for the loads and load arrangements should be taken according to EN 1991 (actions) and EN 1991-8 (earthquakes). Finally, models for verification and specification of material properties can be found in EN 1992-1999.

The general formulation for the partial factor limit state requirement is given by

$$g(X_{d1}, X_{d2}, \ldots) > 0 \qquad \text{1}$$

where $g(\ldots)$ is the limit state or performance function. In EN 1990 this is elaborated as $g = \mathbf{R} - E$, where $\mathbf{R}$ represents the resistance and E the load effect. In this paper the general formulation is followed. Negative values of $g(X)$ correspond to failure and positive values of $g(X)$ to success. The variables X_1, X_2 are the basic random variables like loads, material properties, geometrical properties and so on. The index 'd' refers to the design values of X_i. In the partial factor design these design values are obtained from

$$X_{di} = \gamma_{Fi}\, S_{ki} \text{ (for loads) and}$$
$$X_{di} = \mathbf{R}_k/\gamma_M \text{ (for resistance)} \qquad \text{2}$$

where S_{ki} is the representative value of load i, R_k is the representative value of the resistance and γ represent the partial factors. In some cases also combination factors ψ may enter the equations, expressing the unlikelihood that all independent loads will be extreme for one structure at the same point in time.

The partial factors in the Eurocode have been determined on the basis of a combination of theory and calibration to existing practice in a number of European countries. A sketch of the mixed background is given in Fig. 1. It should be noted that the values in the Eurocodes are recommended values only: every member state has the right to change these values if the need is felt.

The formal link between partial factors on the one hand and probabilistic methods on the other, can be established through the well-established theory of code calibration. The most sophisticated method is to select the set of partial factors in such a way that the sum of weighed deviations forming the target reliability (for a representative set of structures) is a

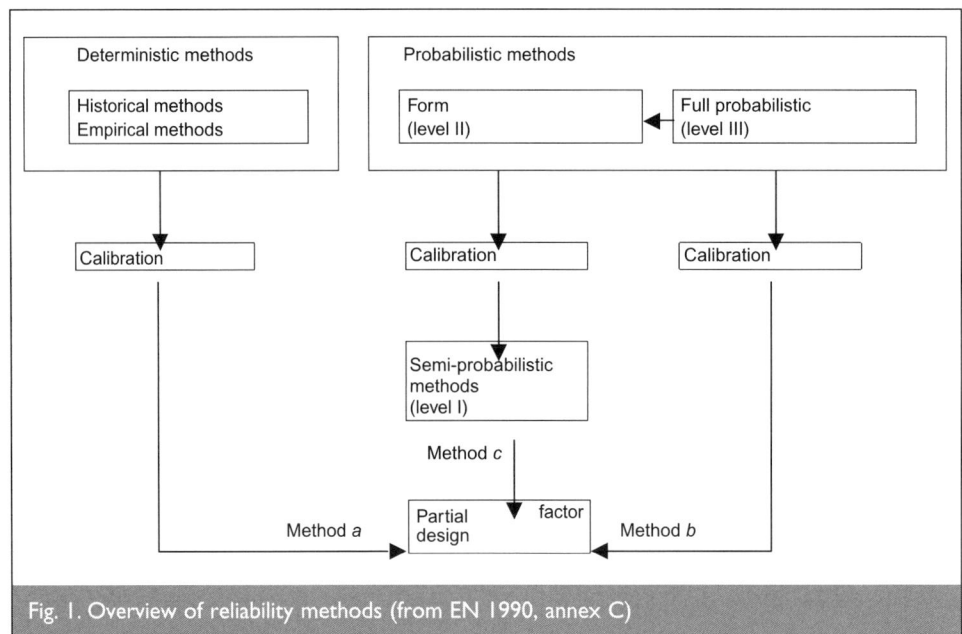

Fig. 1. Overview of reliability methods (from EN 1990, annex C)

minimum.[8,9] A simplified method is the following.[10] For a random variable X (load or resistance) the partial factor is given by

$$3 \qquad \gamma = X_d / X_k$$

In the case of a normal distribution the following expressions hold

$$4 \qquad X_d = \mu(1 - a\beta V) \text{ and } X_k = \mu(1 - kV)$$

where $V \; (= \sigma/\mu)$ is the coefficient of variation, k corresponds to the fractile of the characteristic value, β is the reliability index and α_i the Form sensitivity coefficient[11] $(0 < \alpha_i < 1)$. If $k = 0$ is assumed for simplicity (often the case for loads as the representative value normally corresponds to a return period equal to the design life time), then

$$5 \qquad \gamma_i = 1 + |\alpha_i|\beta V_i$$

Equation (5) is simple and appealing: the partial factor should be increased for a higher influence of the variable, a higher level of reliability and a higher variability. This is a very strong and logical concept. Important is the fact that sensitivity and variability enter explicitly the design equations.

The coefficient of variation V in equation (4) should follow from observations in nature or in the laboratory. The coefficient should also include the uncertainties owing to the limited amount of data (statistical uncertainty), measurement errors and the lack of accuracy owing to schematisation in the calculations (model uncertainty). The index β is related to the level of reliability that is required; roughly, the following relation might serve as a useful approximation for $1 < \beta < 4$

$$6 \qquad P_F \approx 10^{-\beta}$$

From the theoretical point of view, this β-value should follow from an optimisation criterion the expected summed up costs of construction and damage should be a minimum.[12] At present, however, such an approach is not followed in general. The standard procedure is to calibrate the reliability level β to current design levels. As long as the current design practice is not obviously unsatisfactory, either from the safety or from the economical point of view, the average reliability is not changed. The values in EN 1990 are given in Table 1. An

Limit state	Target reliability index*	
	1-year period	50-year period
Ultimate	4, 7	3, 8
Fatigue	—	1, 5 to 3, 8†
Serviceability (irreversible)	2, 9	1, 5

* See annex B
† Depends on degree of inspectability, reparability and damage tolerance

Table 1. Target reliability index β for class CC2 structural members (from EN 1990, annex C)

interesting consequence of the calibration principle is that the choice of the probabilistic models (type of distribution, coefficient of variation) seems to be less critical. For instance, if all coefficients of variation are divided by a factor of 2, calibration will lead to reliability indices β that are twice as high. Then, with the higher β and the lower Vs, the same values are found for the partial factors. This gives a comfortable feeling: it is not necessary to know the statistical models very accurately in order to use this method.

The values in Table 1 hold for the so-called consequences classes CC2. For classes CC1 and CC3 (see Table 2 for definitions) the numbers go up or down by 0·5. Translated into values of partial factors, this means a factor of 1·1 or 0·9 on the load factors.

Finally, the coefficient α in equation (5) is the sensitivity coefficient. This coefficient should, in principle, be derived from probabilistic Form calculations. As the value of a partial factor should hold for a large set of design situations, the value of α may be conceived as an 'average'. In the course of time, certain patterns in the average α-values have been discovered; values according to ISO-2394 and EN 1990 are presented in Table 3.

The dominant value can be found by giving all variables the dominant status one after the other and looking for the combination that governs the design.

Table 3 is the core of the difference between probabilistic design and a probability-based partial factor design: in the probability-based partial factor design the α-factors are fixed in some general conservative way. In all other aspects, the two methods can be considered as equivalent: one may use the same models, the same statistical information and the same

Consequences class	Description	Examples of buildings and civil engineering works
CC3	High consequence for loss of human life, or economic, social or environmental consequences very great	Grandstands, public buildings where consequences of failure are high
CC2	Medium consequence for loss of human life; economic, social or environmental consequences considerable	Residential and office buildings, public buildings where consequences of failure are medium
CC1	Low consequence for loss of human life, *and* economic, social or environmental consequences small or negligible	Agricultural buildings where people do not normally enter (e.g. storage buildings), greenhouses

Table 2. Definition of consequences classes (from EN 1990, Annex B)

	Load	Resistance
Dominant variable	$\alpha_i = 0.70$	$\alpha_i = 0.80$
All other variables	$\alpha_i = 0.28$	$\alpha_i = 0.32$

Table 3. Standardised alpha-values according to ISO 2394

target reliability (see Fig. 1). From this point of view, the statement that 'a deterministic procedure has been followed instead of a probabilistic one because of a lack of statistical data' is not consistent. If there are no data to carry one procedure, it is impossible to carry out the other.

Consider as an example the combination of self-weight G and wind loading W, with $V(G) = 0.10$ and $V(W) = 0.30$ respectively. Applying equation (4) from the main text and target β equal to 3.8 yields the following.

Combination 'G dominant':

$$\gamma_G = 1 + 0.70 \times 3.8 \times 0.10 = 1.27$$
$$\gamma_W = 1 + 0.28 \times 3.8 \times 0.30 = 1.33$$

Combination 'W dominant':

$$\gamma_G = 1 + 0.28 \times 3.8 \times 0.10 = 1.11$$
$$\gamma_W = 1 + 0.70 \times 3.8 \times 0.30 = 1.82$$

For comparison, in EN 1990, one may find $\gamma = 1.35$ for the case of a dominant permanent load and (if one adopts the combinations 6·10a and 6·10b) $\xi\gamma = 0.85 \times 1.35 = 1.15$ for the case where the permanent load is non-dominant. The partial factor for the wind in the Eurocode is 1·5; the above result indicates that for wind loading the Eurocode is less safe than is claimed. From an economical point of view this may be justifiable. Note that the combination value 1·33 for wind combination value would lead to $\psi_0 = 1.33/1.82 = 0.74$. The Eurocode gives 0·60. It should be noted that values may change and calculations become more complex if one deals with the combination of two fluctuating loads, such as wind and snow. Actually the Eurocode does not make such a distinction. In those cases the combination values ψ_0 in the Eurocode are conservative.

4. HAZARDS AND ACCIDENTAL ACTIONS

Accidental actions are defined as actions that seldom occur, but which have a very high impact. One may think of fire, earthquake, impact, explosion and so on. Another type of category is error, misuse or negligence, leading to a weakening of the structure. The design against hazards and other accidental actions is different from design against normal loads. In many cases it is impossible or uneconomic to counteract the loads on the same level of safety. It is therefore better to accept some degree of local damage, but to take measures such that local damage remains local and does not act as an initiation point for progressive collapse. Formally the EN 1990 requires that

A structure shall be designed and executed in such a way that it will

not be damaged by events such as explosion, impact, and the consequences of human errors, to an extent disproportionate to the original cause.

The design with respect to earthquakes is further elaborated in EN 1998. Design with respect to fire is outlined EN 1991-2 and EN 199X-2. The loading specification for impact and natural gas explosions is presented in EN 1991-1-7; this code also presents a strategy against more general and unidentified causes of local damage.

In order to mitigate the risk, structural as well as non-structural measures may be adopted. In principle the Eurocode mentions measures, for example the prevention of hazards, protection or structural elements, reduction of consequences after failure, design for ductility and redundancy and design for sufficient resistance of key elements. Design for accidental design situations needs to be included primarily for structures for which a collapse may have particularly large consequences in terms of injury to humans, damage to the environment or economic losses for society. In practice this means that Eurocode 1, part 1-7 accepts the principle of safety differentiation. In fact the same three different safety categories or consequences classes exist as distinguished earlier (see Table 2). For these categories different methods of analysis and different levels of reliability are accepted. For CC1 there are no requirements, for CC2 there are primarily prescriptive tying measures to achieve sufficient ductility and redundancy and for CC3 there is the advanced risk analysis, including advanced non-linear dynamic structural analysis where relevant.

Annex A of EN 1991-1-7 gives actual guidance for the design of buildings to sustain an extent of localised failure from an unidentified cause without disproportionate collapse.[13] These rules are primarily based on existing rules in UK codes of practice, developed after 1968 following the partial collapse of a 22-storey block of flats in Ronan Point (east London) caused by an ordinary gas explosion. The key conclusion reported at the inquiry[14] was that if the building had been 'as appropriate to a monolithic structure as possible' then it would probably have remained stable, despite having suffered localised damage from the abnormal load of the gas explosion. The UK government was accordingly advised to include provisions within the Building Regulations for dealing with progressive collapse. These rules have proved satisfactory over the past three decades. Their efficacy was claimed to be demonstrated during the IRA bomb attacks that occurred in the City of London in 1992 and 1993. Although the rules were not intended to safeguard buildings against terrorist attack, the damage sustained by those buildings close to the seat of the explosions that were designed to meet the regulatory requirement relating to disproportionate collapse was found to be far less compared with other buildings that were subjected to a similar level of abuse.[15]

For consequences class 3, EN 1991 part 1-7 recommends a risk analysis. Guidance for the planning and execution is given in annex B. The recommended steps are

(a) definition of scope and limitations
(b) qualitative risk analysis (inventory and description)
(c) quantitative risk analysis (modelling and calculations)

(d) risk evaluation and mitigation measures
(e) risk communication.

The depth and complexity should be dictated by the problem at hand. Risk analysis in a rigorous form including extensive statistical analyses will be used only in special cases. In many cases a qualitative analysis of risks and envisaged countermeasures could be sufficient. The actual assessment may often be made by comparison with known structures. In the case of a formal quantitative risk analysis, the following formula for the risk evaluation is presented

$$7 \quad R = \sum_{i=1}^{N_H} p(H_i) \sum_{j}^{N_D} \sum_{k=1}^{N_S} p(D_j|H_i)\, p(S_k|D_j) C(S_k)$$

where $C(S_k)$ are the consequences of adverse state k, $P(H_i)$ is the probability of occurrence of hazard i, $P(D_j|H_i)$ is the conditional probability of damage state j given occurrence of hazard i and $P(S_k|D_j)$ is the conditional probability of the adverse state k given state j. In this formula three basic steps of analysis can be distinguished (see Fig. 2).

Step (a): assessment of hazards with their intensities H_i
Step (b): assessment of different states of damage D_j
Step (c): assessment of the occurrence of the adverse states S_k and consequence(s) $C(S_k)$.

The procedure assumes that the structure is subjected to N_H different hazards, that those hazards may damage the structure in N_D different ways (dependent on the considered hazard) and that the performance of the damaged structure can be discretised into N_S adverse states. The assessment of consequences should be made in terms of injury to humans, damage to the environment, or large economic losses for society. Guidance will be given in the Joint Committee of Structural Safety document on risk analysis, which is under development.

Note that equation (7) is not only valid for rare and incidental loads (such as fire, impact and earthquake), but could also be used to deal with (extreme values of) ordinary loads such as wind and snow. Note further that the probabilities $P(S_k|D_j)$ and $C(S_k)$ can be dependent on time (e.g. in case of fire and evacuation respectively) and the overall risk should be assessed and compared to acceptable risks accordingly.

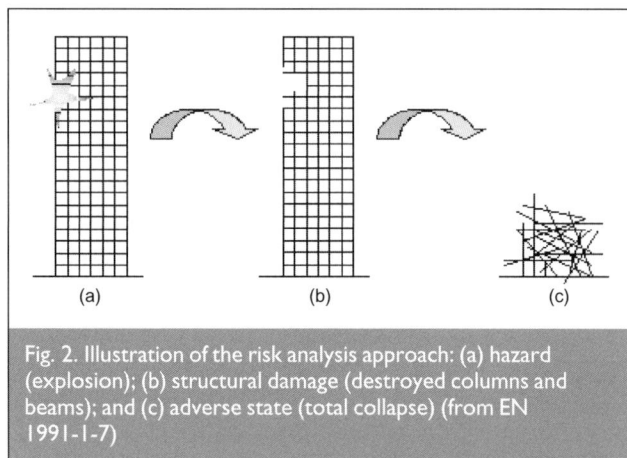

Fig. 2. Illustration of the risk analysis approach: (a) hazard (explosion); (b) structural damage (destroyed columns and beams); and (c) adverse state (total collapse) (from EN 1991-1-7)

An area that typically is left to the member states is the issue of the risk acceptance criteria. Even annex B gives only limited guidance. Basically, the Alarp (as low as reasonably practicable) principle is mentioned. This means that, apart from lower bounds for the individual and socially acceptable risk levels, an economical optimisation is recommended. When evaluating the risk of a certain period of time related to the failure event S on the basis of the consequences C, a discount rate r should be taken into account.

The approach mentioned above may be less suitable for unforeseeable hazards or hazards that are difficult to model (design or construction errors, unexpected deterioration, etc.). For those cases more global damage-tolerant design strategies have been developed, such as the classical requirements of sufficient ductility and tying of elements. A specific approach, in this respect, is the consideration of the situation that a structural member (beam, column) has been damaged, by whatever event, to such an extent that the member has lost its normal load-bearing capacity. For the remaining part of the structure it is then required that, for a relatively short period of time (defined as the repair period T), the structure can withstand the 'normal' loads with some prescribed reliability

$$8 \quad P(R < E \text{ in } T \,|\, \text{one element (partly) damaged}) < p_{\text{target}}$$

The target reliability depends on the normal safety target for the building, the period under consideration (hours, days or months) and the probability that the element under consideration is removed (by causes other than those already considered in design).

5. REFLECTION
As described above, the Eurocode gives a number of tools for dealing with adverse load combinations, the occurrence of accidental loads and the mitigation of the effects for local errors. One should, however, bear in mind that this is only part of the game. Most failures in practice occur despite the existance of a code and despite its application. The use of the code should be embedded in a careful strategy for managing the risks related to structural failure. It is interesting to see that the Eurocode itself recognises this fact. In clause 2.2.1 of EN 1990, it is stated that reliability shall be achieved by (a) a design which is in accordance with the Eurocode and (b) by appropriate execution and quality measures. In annex B this issue is elaborated and a set of different levels of checking procedures and design supervision is presented. The level to be chosen, of course, depends on the consequences class at hand. Aside from those procedures, however, it should be noted that safety management is a matter of mentality. The current author has seen many reliability calculations and risk analyses, typically done for the sake of the record only. They are almost useless.

6. CLOSURE
This paper has summarised the treatment of risk and reliability matters in the newly developed suite of Eurocodes. These standards are based on the partial factor method, which nowadays is a generally accepted design method for building and civil engineering structures. For most engineers the general philosophy that safety factors depend on the type of the load

and on the limit state under consideration makes sense. However, the background, in particular the reliability aspect, seems to be vague and is understood only by a handful of specialists. Even a general notion and feeling has not been developed in the mind of most designers. This is a pity: reliability aspects are as essential for the design as are the mechanical aspects, but the knowledge and appreciation of most engineers is quite unbalanced.

In some cases, of course, the designer may become more involved with statistical aspects. Such exceptions may be when design by testing is carried out, when a risk analysis is to be performed for special structures or some new type of structural solution is considered. Examples are the London Eye[16] or the Maeslandt storm surge barrier in the Netherlands.[17] In such cases the help of experts will be unavoidable.

Another worthwhile point to mention is that the partial factors, characteristic values, PSI-factors and other parameters in the EC may have a probabilistic background, but that the background is heavily mixed with a historical background. In some cases the profession is working hard to improve this fact. The main example is the new Fédération International du Béton model code on the concrete durability design. Other issues that might be developed in the near future are inspection and maintenance, system behaviour and the assessment of existing structures.

Finally, the current paper has emphasised the fact that safety is not a matter of numbers alone: it requires a proper safety management from the start of the design to decommissioning at the end of the structural life.

ACKNOWLEDGEMENT

Permission to reproduce extracts from British Standards is granted by BSI. British Standards can be obtained in PDF or hard copy formats from the BSI online shop: www.bsigroup.com/shop or by contacting BSI Customer Services for hard copies only. Tel: +44 208 996 9001; e-mail: cservices@bsigroup.com.

REFERENCES

1. INTERNATIONAL ORGANISATION FOR STANDARDISATION. *ISO/TC98 General Principles on Reliability for Structures. Revision of IS 2394*. ISO, Geneva, 1994.
2. JOINT COMMITTEE ON STRUCTURAL SAFETY. *Probabilistic Model Code*. JCSS, Zurich, 2001. See http://www.jcss.ethz.ch/
3. DITLEVSEN O. and MADSEN H. O. *Proposal for a Code for the Direct Use of Reliabilty Methods in Structural Design.*

International Association for Bridge and Structural Engineering (IABSE) publication, Zurich, 1989.
4. LÜCHINGER P. Basis of design serviceability aspects. *Proceedings of an IABSE Conference on Basis of Design and Actions on Structures*, Delft, 1996.
5. DITLEVSEN O. Uncertainty and structural reliability, Hocus Pocus or objective modelling? *Proceedings of the 10th International Ship Structures Congress*, Lyngby, 1988.
6. DITLEVSEN O. Decision modelling and acceptance criteria. *Structural Safety*, 2003, **25**, No. 2, 165–191.
7. ELMS D. G. and TURKSTRA E. G. A critique of reliability theory (BLOCKLEY D. I. (ed.)). In *Engineering Safety*. McGraw Hill, New York, 1992.
8. VROUWENVELDER A. and SIEMES A. Probabilistic calibration procedure for the derivation of partial safety factors for the Netherlands building codes. *Heron,* 1987, **32**, No. 4, 9–29.
9. FABER M. H. and SØRENSEN J. D. Reliability based code calibration: the JCSS approach. *Proceedings of ICASP9*, San Fransisco, 2003, **2**.
10. VROUWENVELDER A. Reliability based code calibration: the use of the JCSS Probabilistic Model Code. *Proceedings of JCSS Workshop on Code Calibration*, Zurich, 21–22 March 2002.
11. MADSEN H. O., KRENK S. and LIND N. C. *Methods of Structural Safety*. Prentice Hall, Englewood Cliffs, New Jersey, 1986.
12. RACKWITZ R. Optimisation: the basis of code making and reliability verification. *Structural Safety*, 2000, **2**, No. 1, 27–60.
13. GULVANESSIAN H. and VROUWENVELDER T. Robustness and the Eurocodes. *Structural Engineering International*, 2006, **16**, No. 2, 167–171.
14. PUGSLEY A. and Saunders O. *Collapse of Flats at Ronan Point, Canning Town*. Ministry of Housing and Local Government, Her Majesty's Stationery Office, London, 1968.
15. INSTITUTION OF STRUCTURAL ENGINEERS. *Safety in Tall Buildings and Other Buildings with Large Occupancy*. IStructE, London, 2002.
16. DIJKSTRA O. D., BERENBAK J., VROUWENVELDER A. W. C. M. and SNIJDER H. H. The British Airways London Eye: design review and special studies. *Proceedings of an International Conference of Safety, Risk and Reliability Trends in Engineering*, Malta. International Association for Bridge and Structural Engineering, Zurich, 2001.
17. FOEKEN F. J., VAN KERSTENS J. G. M., VAN MANEN S. E. and VROUWENVELDER A. C. W. M. Probabilistic design of the steel structure of the 'Nieuwe Waterweg' storm surge barrier. *Proceedings of the 4th International Offshore and Polar Engineering Conference, Osaka, Japan*, 1994, **483**, Vol. IV.

What do you think?

To comment on this paper, please email up to 500 words to the editor at journals@ice.org.uk

Proceedings journals rely entirely on contributions sent in by civil engineers and related professionals, academics and students. Papers should be 2000–5000 words long, with adequate illustrations and references. Please visit www.thomastelford.com/journals for author guidelines and further details.

Proceedings of the Institution of
Civil Engineers
Structures & Buildings 161
August 2008 Issue SB4
Pages 215–218
doi: 10.1680/stbu.2008.161.4.215

Paper 800017
Received 05/02/2008
Accepted 17/03/2008

Keywords: structural frameworks/
risk & probability analysis

David A. Nethercot
Professor, Department of
Civil and Environmental
Engineering, Imperial College
London, UK

Reliability, responsibility and reality in structural engineering

D. A. Nethercot OBE, PhD, DSc, FREng, FICE, FIStructE, FCGI

For a person to be described as reliable or for a service to be regarded as something that can be relied upon is an easily recognised and highly desirable description. Yet when reliability is mentioned within the structural engineering community, it provokes decidedly mixed reactions. For the enthusiasts—found almost entirely within the academic community—it is a key tool in the quest to understand and describe ever more challenging aspects of structural behaviour, but to the practitioner it is largely an irrelevance: something that belongs elsewhere with no real place in their everyday activities. They may have read something about 'the improved levels of reliability' assured by newly introduced codes of practice, recall that it is somehow related to risk or note increased use of the word in project specification documents. Why this difference in attitude? What are the reasons? Can the divide be bridged? Does it matter? This paper presents a largely personal view on these issues.

I. INTRODUCTION
The objective of structural design may be thought of as the generation of the information necessary for the construction of safe, economic, functional and attractive structures. Achieving this requires careful use of engineering judgement since many of the issues will not have a single solution; many design objectives may well exhibit an element of contradiction and decisions will normally have to be made on the basis of less than certain data, imprecise quantitative analysis and features that can only be considered in a qualitative fashion. How does the structural engineer reconcile the seemingly impossible challenges? How does he/she decide on the merits of various tradeoffs? What are reasonable acceptance criteria? Taking a single aspect of the objectives–safety; how safe should the structure be and how certain should the designer be that this level of safety has been achieved?

Questions of this sort clearly cannot be answered solely by the use of theories based on the principles of mechanics. Other factors must be brought into the process. It is here that the principles of reliability theory, applied in a pragmatic and appropriate fashion, can assist as they represent a structured way of dealing with variability and uncertainty of the type inevitably present in any structural design. Proper recourse to suitable processes should assist in ordering structuring thinking, identifying key influences, resolving potential

contradictions, making sound decisions and demonstrating the integrity of the final answer.

This paper is not a discourse on reliability theory; others in this issue address various aspects of that topic. Rather, it is an essay on what should be our approach to various features of structural design, intended to expose both the inherent complexities and the need for intelligent simplifications based on sound principles. Reliability theory provides several ways to assist with this.

2. THE REALITIES OF STRUCTURAL DESIGN
Although engineering makes use of the 'scientific method', it is fundamentally different from science in that it requires the devising of a workable solution, no matter how complex the problem posed. Science, on the other hand, is generally regarded as being a search for ever more refined descriptions of phenomena. Of course, being able to represent certain aspects of an engineering problem in a scientifically more rigorous fashion has a place within the advance of engineering but it is the totality of the answer and its overall relationship with the full set of problem constraints and devising the solution that 'works' that is the essence of engineering.

Thus in structural engineering it is the ability to identify and define all the problem constraints and then to devise the solution that best satisfies all of these in the most balanced way, that is, recognising that different constraints need to be considered to different levels of importance, that is the key objective. Given the complexity of real structural engineering problems, there will always be an element of uncertainty in the extent to which the available theories and design procedures fully represent all aspects of the physical problem as well as the accuracy and dependability of all the input data and physical descriptions that feed into the process, that is, the problem will not be deterministic in the sense that values used in calculations can be relied on with certainty but will represent estimates or ranges. Moreover, the certainty of different items of input data will vary considerably, for example Young's modulus for steel hardly varies so it can almost be regarded as a deterministic quantity, while the practice of selling structural steel on the basis of guaranteed minimum strength means that, although getting material below specification is extremely unlikely, finding that the material is (say) 30% over the specified yield strength is not uncommon. This may not be a benefit for those situations in which some elements within a system that control failure and have been selected so as to exhibit ductility upon yielding, thereby ensuring a benign form of failure, are actually stronger than expected, with the result that other (less suitable) components now control the failure.

One often-quoted, although not altogether serious, statement on this is

> Structural Engineering is the art of moulding materials we do not wholly understand into shapes that we cannot precisely analyse, so as to withstand forces we cannot really assess, in such a way that the community at large has no reason to suspect the extent of our ignorance.
> (Chairman's Address to the Scottish Branch of IStructE by Dr A. R. Dykes, 1978.)

Clearly this is a caricature but as with all such representations it contains more than a grain of truth.

Turning to the challenges of structural design, these include the following.

(a) Loading: the various possible types need to be identified, their likely magnitudes assessed, the extent to which they are likely to act in (various) combinations decided and representations suitable for inclusion in calculations devised, for example wind load is intermittent, dynamic, variable in direction and constantly changing over time– how should this be represented for design purposes?

(b) Materials: key structural properties need to be defined and the appropriate values assigned, perhaps a more comprehensive representation of some aspects of behaviour devised, for example the relationship between temperature, time and strain if behaviour at elevated temperature is a factor, and material behaviour integrated into the processes of determining structural response.

(c) Analysis: some way of assessing the consequences of the application of loads to the structure, that is the resulting deflections, internal member forces, stresses, and so on, needs to be devised in the form of a model that will permit suitably representative quantitative assessments to be made.

It should be clear that each of these three key steps involves some degree of variability and uncertainty. Loading cannot be modelled exactly, material behaviour will represent the essence but not all features of the performance, and analysis will be based on a set of simplifications and assumptions in order to make it tractable. How can all of this be reconciled with the public's legitimate expectation of adequate safety?

3. THE BASES OF STRUCTURAL DESIGN

Structural design has progressed through a number of different formats: permissible stress, limit states, performance based and so on. Each represents an attempt to more clearly define and satisfy the various objectives. Taking the currently most widely used, limit-states design (LSD), as an example, it is generally accepted that this concentrates on both the ultimate and the serviceability limit states. There is, however, growing realisation that a third condition–often termed 'extreme events'–also needs attention. In this respect, it is informative to look at the procedures used by the seismic design community

that exhibit greater recognition of a hierarchy of different conditions. Let us consider the three limit states in turn (serviceability, ultimate and extreme) and the extent to which all the important information can be unequivocally defined and specified.

4. SERVICEABILITY

Taking as an example the requirement for acceptable deflections under (static) loads, it is generally the case that calculations are undertaken on an elastic basis using a scaled down version of the ultimate loads. Both assumptions deserve some examination.

For steel construction, it has long been recognised[1] that residual stresses exist within members and that these lead to early yielding. Similarly, studies for composite steel/concrete beams[2] indicate irreversible deflections at a much earlier stage than would be supposed through the simple use of current code rules. All structures are, of course, susceptible to the effects of lack of fit causing differences in the internal stress distribution from that calculated on the basis of perfection. Similarly, common assumptions regarding the behaviour of connections– pinned or rigid–fail to take into account the semi-rigid and partial strength nature of the majority of practical arrangements. Materials commonly thought to exhibit a simple linear relationship between stress and strain often demonstrate this only for a very limited range of behaviour, for example structural steels produced by modern production routes generally do not exhibit the elastic-perfectly plastic behaviour that forms the basis for many design treatments. Similar lists of departures from the conventional set of assumptions (used in the analysis) could be assembled for all the commonly used structural materials.

Considering the loading side, the presumption that serviceability loading is simply the ultimate case with a load factor of unity makes no allowance for differences between the regular and severe conditions. Moreover, when extended to load combinations, the balance between different actions in the regular service conditions are likely to be quite different from those of the severe ultimate conditions.

A third source of uncertainty is the actual criteria used for acceptance. These are largely rather unscientific,[3] being based on a history of satisfactory performance in the past, that is their use has generally resulted in the absence of problems. Forms of construction and approaches to design change, however, so there must be a question mark against continuing reliance on a historic consensus. Moreover, such an approach throws up several inconstancies, for example drift calculations for sway frames are normally conducted on the assumption that the beam to column joints function as rigid, but a more realistic approach would include some measure of flexibility, leading to larger displacements and thus making it more difficult to satisfy the design requirement. It is, of course, illogical for a more precise representation of behaviour to be penalised by having to work to criteria that implicitly make some allowance for an effect that is now being included explicitly.

Designers, or more likely, those responsible for the production of design rules and specifications, need some framework within which to reconcile these seemingly unquantifiable issues.

5. ULTIMATE LIMIT STATES

Design for the ultimate limit state requires that an appropriate representation of the structural response up to the maximum load be available. Whatever method is used, it will, of course, include a number of assumptions and limitations. For example, simple plastic theory neglects elastic deflections and makes very simple assumptions regarding member behaviour that, in turn, rely on approximations to material behaviour. It does, however, have the great attraction that for the range of structures for which it is suitable it provides good approximations with a minimum of calculation.

How should the use of such a method be reconciled with an alternative that more correctly allows for the true behaviour? If such a situation is not recognised, what motivation exists for adopting a more rigorous approach?

Recognising the increasing tendency for structures to become more slender and thus for instability effects to be increasingly likely to exert a significant influence on behaviour, many structural codes of practice have incorporated provisions for the conduct of some form of second-order analysis of overall system response. Since undertaking any form of non-linear analysis is, in the view of the writer, an order of magnitude more challenging than relying on linear analysis, an appropriate basis for deciding when this is required and how the findings should be used *vis a vis* those from an alternative provision for adapting the findings of a linear analysis to approximate for the non-linear effects is needed. How should these two alternative approaches be compared in a meaningful way and what extra margins are necessary when using the simpler but less accurate approach so as to ensure complementary levels of safety between the two? Once again recourse to reliability theory provides a consistent framework for doing this.

6. EXTREME EVENTS

Events such as progressive collapse resulting from terrorist action are, by their very nature, expected to occur only rarely. These 'low probability but high consequences events' are the most difficult to address. There is never likely to be a significant bank of data available against which to compare design treatments, the consequences of over provision are likely to be significant in practical and economic terms, while under provision runs the risk that if/when such events occur, public opinion may well take the view that the professionals have been negligent in their duty.

At the present time this situation is rendered more difficult by the relative lack of soundly based design provisions[4] to limit the likelihood of progressive collapse following the occurrence of limited damage to a structure. Most of the currently used design approaches are prescriptive in nature and might best be summed up as following the 'if this is done then behaviour should be better than if it is not'. There is an absence of linkage between cause and effect, of the ability to make quantitatively based comparisons between alternative arrangements or, indeed, of a real understanding of the complex interplay of the physical phenomena involved.

Given the complexity of the subject any attempt at a reasonably consistent analysis would need to recognise

(a) events take place over a short timescale and the actual failure is therefore dynamic in nature
(b) it involves gross deformations, generating large strains, leading to inelastic material behaviour as well as change of geometry effects
(c) failure essentially corresponds to an inability of the structure in its damaged state to adopt a position of equilibrium without separation at key locations.

Ideally the complexity of the design approach adopted should accord with the importance of the structure and/or the severity of the consequences, that is simple approaches for routine structures and/or situations for which progressive collapse might not be considered quite so catastrophic but more rigorous approaches for major structures and/or situations with the potential for very severe consequences.

Although this topic has yet to reach the level of maturity of design for the serviceability and ultimate limit states, it is of interest to note that reliability concepts are already being advanced[5] as being necessary for the development of appropriate design frameworks.

7. CODES OF PRACTICE

Structural codes of practice have been an important feature in the work of structural engineers for many decades. It is therefore inevitable that each is subject to a process of review and updating—supposedly at 10 year intervals according to BSI rules but more or less frequently in reality. As an example, Table 1 lists the various amendments and revisions to the UK's code for the structural use of steel in buildings BS 5950 and its predecessor BS 449. While the former is limit states based, the latter was, of course, a permissible stress code. Those responsible for their production were required to make decisions on the values of the various safety factors to be used as design rules, forms of construction and the material itself (that is properties of structural steel), evolved with time.

Clearly some elements of judgement were required, for example, if a newly proposed design rule could clearly be shown to provide a better fit to all the relevant test and numerical data for the topic under consideration there was a

BS 449 history for part 1
1932: 1st edition
1935, 1937, 1948, 1959: Revisions
1969: 5th (metric) edition
1970, 1973, 1975, 1984, 1988, 1989: amendments
1990: New version
1995: Amendment 9

BS 5950 history for part 1
1985: 1st edition
1990: Revised
2000: Revised

Table 1. Typical code revisions

clear case for reducing any margin previously incorporated to cover uncertainty. But how could this be done consistently for all parts of the code? Some form of calibration of the results from the new rules against predictions from the equivalent existing rules (taken to be satisfactory on the basis of hitherto acceptable performance) was required. Once again reliability theory provided the basis for conducting such studies across a variety of phenomena, using data of different extents and quality, in a consistent fashion.

8. CONCLUSIONS

Structural engineering inherently relies on the use of reliability concepts in ordering its thinking and in supporting the processes it uses. Normally much, if not all, of this is hidden within the actual application rules, analysis techniques, guidance material and so on. Its use has, however, been essential in placing these on a consistent basis. Since the answer of most structural engineers to the questions

(*a*) how safe is your structure?
(*b*) how sure are you of your answer?

is likely to be along the lines of 'I can't actually say but I have followed the best available approaches in my design', we should be grateful for the presence of reliability theory as an essential component in supporting our work.

REFERENCES

1. KETTER R. L. The influence of residual stresses on the strength of structural members. *Welding Research Council Bulletin No. 44*, 1958.
2. NETHERCOT D. A., LI T. Q. and AHMED B. A. Plasticity of composite beams at serviceability limit state. *The Structural Engineer*, 1998, **76**, No. 15, 289–293.
3. SAIDINI M. and NETHERCOT D. A. Serviceability deflections and displacements in steel framed structures; Code requirements and their bases. *IABSE Proceedings*, 1995, **5**, No. 2, 110–113.
4. NETHERCOT D. A., IZZUDDIN B. A., ELGHAZOULI A. E. and VLASSIS A. G. Aligning progressive collapse with conventional structure design. *ICASS 07* (LIEW R. and CHOO Y. S. (eds)), Vol. 1, Research Publishing, 2008, pp. 1–21.
5. STAROSSEK U. Disproportionate collapse: a pragmatic approach. *Proceedings of the Institution of Civil Engineers, Structures and Buildings*, 2007, **160**, No. 6, 317–325.

What do you think?
To comment on this paper, please email up to 500 words to the editor at journals@ice.org.uk

Proceedings journals rely entirely on contributions sent in by civil engineers and related professionals, academics and students. Papers should be 2000–5000 words long, with adequate illustrations and references. Please visit www.thomastelford.com/journals for author guidelines and further details.

Proceedings of the Institution of
Civil Engineers
Structures & Buildings 161
August 2008 Issue SB4
Pages 219–230
doi: 10.1680/stbu.2008.161.4.219

Paper 800010
Received 21/01/2008
Accepted 23/04/2008

Keywords: safety & hazards

C. B. Brown
Professor Emeritus of Civil Engineering,
University of Washington, Seattle, WA, USA
and Courtesy Professor of Civil Engineering,
Oregon State University, Corvallis, OR, USA

D. G. Elms
Emeritus Professor of Civil Engineering,
University of Canterbury, Christchurch, New
Zealand

R. E. Melchers
Professor of Civil Engineering, University of
Newcastle, New South Wales, Australia

Assessing and achieving structural safety

C. B. Brown AKC, PhD, PEng, DistMASCE, D. G. Elms PhD, FRSNZ, DistFIPENZ and R. E. Melchers PhD, FICE,
FIEAust, MASCE, FTSE

Structures seldom fail, so it is reasonable to conclude that they are 'safe' and that this state was attained in the processes of designing and building them. The process of achieving structural safety and the process of assessing whether structural safety has actually been achieved for a given structure are, however, two very different matters. This is the essence of this paper and the issues are discussed in terms of the four related topics of responsibility, failure, uncertainty and decision. The first deals with the domain of responsibility of structural engineers: what it is and what it should be and how this relates to structural failure. Four sources of failure are distinguished: technical, process errors, technical ignorance and various non-technical matters. The relative contributions of these are then discussed. Recent codes of practice have focused narrowly on technical uncertainty and a notional probability of failure, with little attempt to incorporate non-technical matters and the ontological uncertainty, which is shown to be the origin of most failures. Finally, the nature of decision making is discussed for both code development and structural design and it is concluded that in both cases the decision process is one of satisficing rather than optimising. The whole process is less rational than is generally supposed and requires a broader view in the outlook and training of structural engineers.

NOTATION

D	domain of consideration (all structures, some structures, one structure and so on), the universal set
E	sub-set of D in which errors occur
F	sub-set of D in which failures occur
FO	observed failures
G*	the actual boundary between F and S.
G	the estimate of G*
K	intersect of F and E
N	notional concerns
p_X	the probability of the occurrence of X
S	subset of D in which failures do not occur
T	failures owing to technical concepts
TU	failures owing to technical errors
U	failures owing to non-technical matters
UT	failures owing to technical ignorance

I. INTRODUCTION

As there are relatively few failures, it can be assumed that in general the safety of engineered structures is achieved. To achieve safety in any proposed structure it is conventional that the structural engineer considers whether a design satisfies the various criteria specified in codes of practice. The process seldom, if ever, involves the explicit consideration of the likely probability of failure of the structure that will result from the design and construction work. The assessment of safety, on the other hand, can be done in a variety of ways, but in a formal setting involves the estimation of the probability of failure for a proposed or an existing structure and ensuring that this probability is sufficiently low against a set of agreed criteria. In practice the probability of failure and the safety criteria may be measured by less formal surrogates. The safety level actually achieved is revealed by the statistics obtained from real data. Such achieved probabilities of failure are usually at least two orders of magnitude larger than the estimated or assessed values.

The present paper discusses the relationship between assessment and achievement. It argues that the relationship between the two is not at all clear, and concludes that structural engineers need a different and broader outlook than their training and employment currently provides. The paper is divided into four issues regarding structural safety. responsibility, failure, uncertainty and decisions. Each issue is considered in a separate section.

It is useful to begin by considering three figures (Figs 1–3). Here 'structure' is used quite generally and may be interpreted to include a structural element, a complete structure or even a group of structures. The regions in Figs 1–3 are more properly interpreted as 'sets' in set theory. The so-called 'universal set' is whatever is being discussed and is the domain D. It includes all reasonable forms, properties and dimensions the structure could have. This means that in the figures it includes sets S and F, that is, D is the union of S and F, with the intersection of S and F being the empty set. S and F are separated by a boundary G*. This is the actual but unknown boundary and G in Fig. 2 is a professional estimate of G*. In a world of complete certainty, as in Fig. 1, any structure defined by a point in S would have a probability of failure, p_F, of zero and any structure in F a value of one. In the real world such certainty does not exist and a diagram in the form of Fig. 2

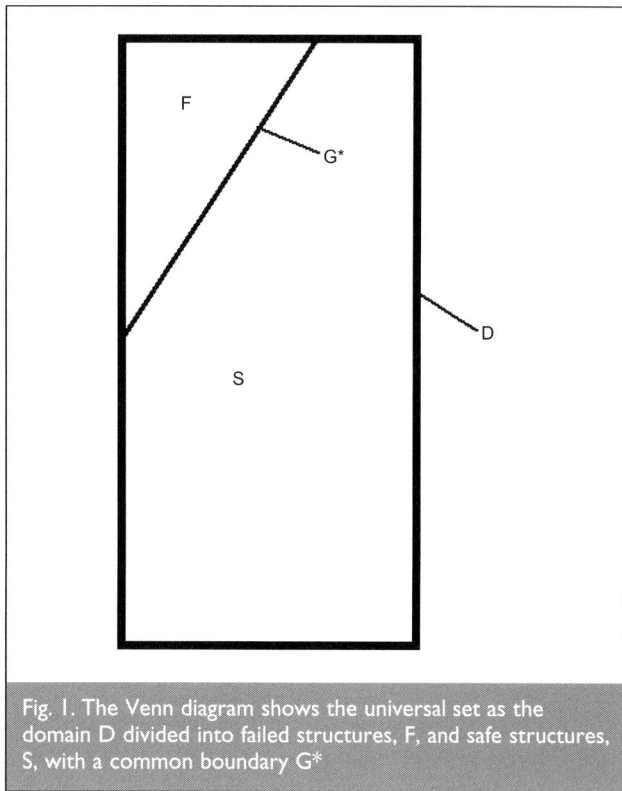

Fig. 1. The Venn diagram shows the universal set as the domain D divided into failed structures, F, and safe structures, S, with a common boundary G*

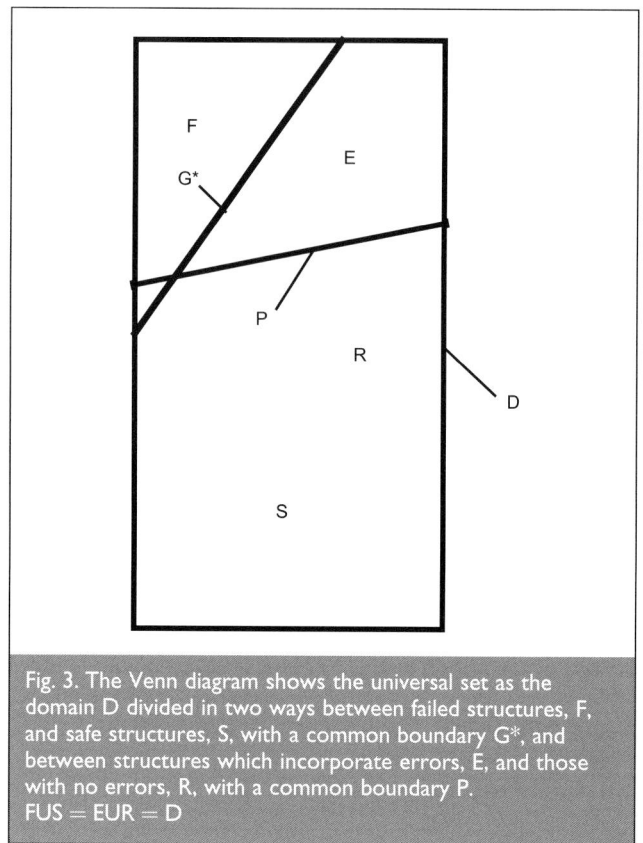

Fig. 2. The Venn diagram shows the universal set as the domain D divided into failed structures, F, and safe structures, S. The actual boundary between F and S is G* and the estimated boundary is G

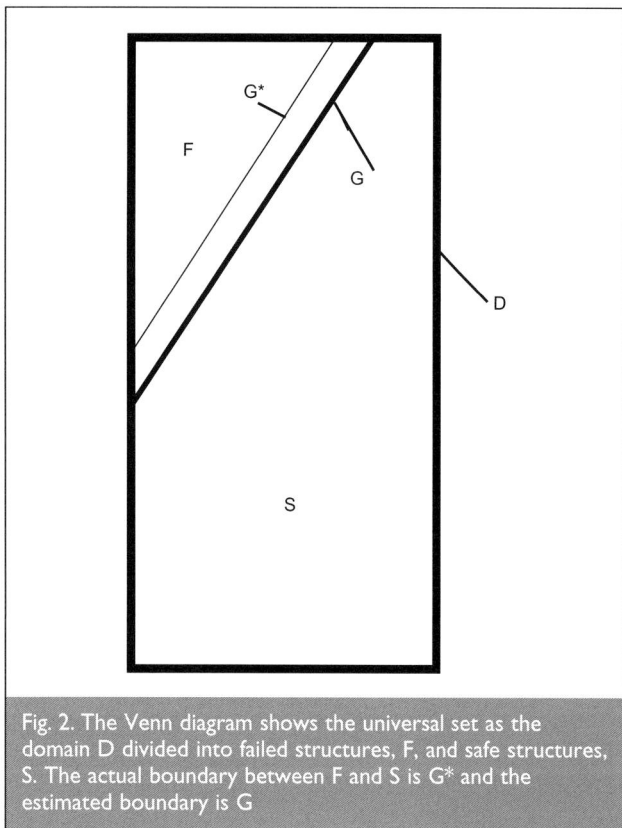

Fig. 3. The Venn diagram shows the universal set as the domain D divided in two ways between failed structures, F, and safe structures, S, with a common boundary G*, and between structures which incorporate errors, E, and those with no errors, R, with a common boundary P. FUS = EUR = D

assessment of structural reliability concerns the analytical procedures that are necessary to estimate not only this value but also the credibility of this value. The achievement is concerned with the success of such assessment and hence level of safety in a real situation. The prescribed boundary, G, in Fig. 2 and the domain of discourse, D, are critical in the assessment. The domain, D, is the collective features being discussed. D can, for example, be the bridges on a transportation system, the bridges of a certain type on the system, one of those bridges, or a member in such a bridge. What is being discussed, that is D, has to be sharply defined.

The domain, D, identifies the professional responsibility level of the decision. This is discussed in detail below in section 2, which considers a hierarchical range of domains that could determine the professional responsibility for the reliability within D. The sources of failure are defined and their combination identified from these considerations. To this stage little quantitative material has been introduced but with these sources of failure available, data corresponding to the sources are considered. Section 3 also includes explanations about errors and their effect on attaining safety. Fig. 3 shows the interplay between errors and safety together with the world of uncertainty exhibited in Fig. 2. This uncertainty includes the boundary between safe and unsafe structures, G, and the incidence of errors resulting in failure, the intersect of F and E. Section 4 considers these matters. The background that has been established provides the basis for decision making. There is a tendency to think of this as design and to start a discussion on design codes. As different domains become the focus, however, it is apparent that decision making is perceived differently at different levels and at different times in the life of a structure. Certainly design, construction and codes have their place, but an understanding of the assessment of

begins to describe the situation. Here the boundary between S and F is uncertain and this uncertainty is described by the probability of failure, p_F, of a structure in D lying between zero and one. The boundary G indicates the professional estimate of the separation between F and S and the location of the feature relative to that boundary is an indication of its reliability. The structural safety and the consequent achievement cannot be undertaken without broadening the discussion. Decisions are not only based on the chances of failure but also the consequences. The consideration of risk and the broadening of the considerations involved in structural safety are taken up in section 5. The four sections on responsibility, failure, uncertainty and decisions provide the background and context for a closing discussion on uncertainty and its relationship to structural reliability.

2. RESPONSIBILITY

Practicing structural design engineers have no difficulty in defining the domains of their work. Most are preparing designs of members and connections where safety is ensured by compliance to the relevant codes of practice. They are supervised within a larger domain where the intentions of the responsible engineer are enforced. That engineer also has responsibilities to satisfy the requirements of the owners. These are transmitted by architects, planners and other non-technical people. It is at this stage that this neat hierarchy of domains often breaks down and domains exist which call for engineers with understanding beyond the technical arena.

The various interacting domains do provide a process that leads to the preparation of bid documents, appointment of contractors, construction and acceptance by the owner. While structural engineers are active in design matters and contract preparation, the owner's interests are protected by the responsible engineers. The domain of the latter requires the understanding of the design, contract documents and the competency of the contractors, all intended to ensure that the structure is completed as designed.

Failures can occur at any stage and conditions can be generated in the design–construct process that can lead to subsequent failure. The concerns with respect to structural safety in this system are largely technical matters with which engineers may be assumed to be comfortable to the extent that their professional education and training has prepared them for this work. There is little doubt concerning the domain of activity of the various engineers and to whom they are responsible. Design–construction appears to be an orderly system in which the various activities seem to nestle like Russian dolls. Undoubtedly in practice many possibilities exist for failure to occur. Some can be associated with bounds in technical understanding having been transgressed and important information such as about material properties not being available. Let the probability of occurrence of such failures be termed p_T.

A different cause of failure is undischarged responsibilities. This can occur through errors in the process. As an example, the flange width was confused with the web width in the design of the temporary structure of the Second Narrows bridge in Vancouver, Canada in 1958 and led to the collapse of the bridge during construction. The evidence to date suggests that the gusset plates were undersized on the Mississippi bridge in Minneapolis which failed in 2007, 40 years after construction. The reason for these failures can be considered technical but would not have happened without error. Herein such failure probabilities are termed p_{TU}. The difficulty in assigning a cause is, however, well illustrated in the case of the

Mississippi bridge. Of course ignorance about technical matters can exist amongst practitioners and lead to technical failures. These are designated p_{UT}. Pugsley[1] pointed out that all technical work is influenced by non-technical factors which he termed 'climates'. They are

- (a) new or unusual materials
- (b) new or unusual methods of construction
- (c) new or unusual types of structures
- (d) experience and organisation of design and construction teams
- (e) research and development background
- (f) industrial climate
- (g) financial climate
- (h) political climate.

Since then the list has been amplified but the point is that the neat, tidy domains associated with technical activities often have to be augmented to include such non-technical features. There are many known cases of structural failure associated with these psychological, organisational and sociological interactions. For instance an examination of 12 structures damaged in earthquakes indicated marked attenuation of safety in eight of them owing to such non-technical interactions.[2] Herein the probability of failure of structures resulting from these non-technical matters is termed p_U.

The broadening of discourse to domains that include disciplines that are unfamiliar to structural engineers can be difficult and yet participation in these extended domains is essential in the assessing and achieving of structural safety. Usually the interactions are with other professionals but the tendency to include all perceived stakeholders in discussions that involve structural matters requires an involvement with lay communities. For example, General Riley issued an edict requiring 'shared vision planning' in the civil arm of the US Corps of Engineers.[3] In this way anyone who claims an interest in future civil engineering work by the Corps is invited to intrude into the planning and activation procedures. This is not a trivial matter as Riley supervises US$5 billion annually of civil engineering work.

An expanded domain of interaction by engineers also brings out some other issues. For example, 'engineers' formal education seldom prepares them for wider roles. Moreover, the strident urging 'to think outside the box' can easily fall on deaf ears because it is not a part of the chosen career path and many employers do not actively encourage it. To deal competently with structural safety, however these broader domains leading to p_U have to be included. The process advocated by Riley would likely occur before the technical design and involve both qualitative and quantitative assessment by structural engineers, other professionals and a lay population. In practice it is this stage that reveals potential 'hidden' or 'forgotten' issues that may cause subsequent safety problems. In a more formal setting the interacting professionals can establish arrangements with definite responsibilities. Such discourses have to be conducted in a systematic manner with conclusions that identify responsibilities. A debate in itself, although often interesting, will not advance the achievement of structural safety. This broader discussion of the fundamental issues of safety is seldom conducted by structural engineering professionals even though there is a wealth of literature on the subject.[4]

These four probabilities of failure can be considered as sub-sets on a Venn diagram where the universal set, F, is associated with the actual probability of failure. T, TU, UT and U have been defined such that the union is F and the intersect the empty set. Under these circumstances they are independent and the union operation then leads to the summation

$$p_F = p_T + p_{TU} + p_{UT} + p_U \qquad (1)$$

This careful statement leading to independence does not attenuate the difficulty in assigning the appropriate sub-set for an actual failure.

The first three terms on the right-hand side of equation (1) represent work that falls under the authority of the responsible engineer (or engineer of record). This appears to be a clear and unambiguous responsibility. The breakdown of the technical part of the safety environment is, however often initiated by Pugsley's climates and other non-technical features. An example of this is the failure of the First Tacoma Narrows bridge in 1940 which initiated an exhaustive technical study of the behaviour of suspension bridges subjected to wind. At first blush the failure would be included under p_T. However the picture changes when the history leading up to the failure is considered.

Funding for the structure was in the hands of the Washington Toll Bridge Authority which authorised the design of a structure. This design featured an 850 m central span with a 7·6 m deep open truss stiffening beam. The span:depth ratio of 112 was typical of most bridges constructed since the Brooklyn bridge. The Authority then applied for monies to the Federal Public Works Administration in Washington, D.C., which agreed to provide 45% of the cost as long as a review board was established to examine the design. As a result, new designs of the super- and substructures were made, which were then constructed.

The superstructure consisted of the same span but was stiffened by two 2·43 m deep plate girders. The span:depth ratio was 350. It was this bridge that failed on November 7, 1940 and led to one of the most complete technical studies in structural engineering. The existing replacement bridge has the same span and a span:depth ratio of 85; the stiffening is with an open truss, 10 m deep.[5] The failure is usually considered technical, but it can also be argued that the bridge was doomed once the original design was abandoned. The category defined by p_U would then be appropriate and the items of financial and political climates on Pugsley's list would have to be considered.

A domain of responsibility exists that is outside the technical and is identified with the climates of Pugsley. This domain requires understanding beyond that of technical matters and is a part of life in other endeavours. Those experiences and methodologies have been studied by Blockley[6] and Henderson and Blockley[7] within the philosophies of Popper and of Kuhn. Melchers[8] has considered these non-technical matters from a systems view. The call is for decision makers to 'think outside the box' but there is as yet no clear, unambiguous responsibility for this extended vision of structural safety.

There is no trouble in believing that the assignments of responsibilities have been correct when things go well. It is only when confronted with failure that difficulties arise. Even when the failure is apparently due to technical reasons, as in the case of the Tacoma Narrows bridge, it may be possible to identify climates that ensured critical decisions that impacted the technical solution. In such circumstances it is not clear whether a T or a UT or a U type failure has occurred. For effective professional progress it is essential that a system exists that ensures that those with responsibility are at least aware of these circumstances while the design–construction operational sequence is underway.

3. FAILURE

The usual practice of design, construction and use anticipates that failures will occur. The codes of practice description of the loads, structural capacities and the boundary, G, in Fig. 2 are refined. There is, however, an abundant literature to show that failures are seldom the result of technical shortcomings. They are dominated by errors in design and construction, professional ignorance, and non-technical matters. Any collection of failure data will therefore tend to be describing p_{TU}, p_{UT} and p_U rather than p_T. A stronger statement can be made[8] that

$$p_{FO} \approx p_U \qquad (2)$$

where p_{FO} is the observed failure rate. The counting of the observed failures is usually arranged as statistics and they depend on the domain being considered. Some obvious domains are bridges, buildings and suchlike categories. Sub-divisions can, however, also be informative. For instance, in the state of Washington five floating bridges have been built of which two have failed. In both cases the loadings surpassed the bridge capacities and the failures can reasonably be thought of as technical. This gloomy picture of $p_{FO} = p_T = 0·4$ has not dimmed the enthusiasm for this structural type inasmuch as a replacement for one of the three extant bridges is being completed which is also a floating bridge. Clearly this type of statistic is not taken too seriously by the owners and draws attention to requirements for the sample counted.

Certainly the numbers involved must be large enough and the sample truly representative of the population. These are largely subjective decisions even though helpful decision schemes can be employed. In the past such a congregation of data into domains has been treated very differently by different statisticians. For instance in the 19th century the administrative statisticians in England had no trouble lumping all of the population together to identify public health concerns, whereas the French administrators baulked at such homogeneity on the premise that each French citizen was unique.[9]

More typical of structural safety is the observation by Fitzroy that the *Beagle* of Darwin fame was one of a class of 107 vessels of which 26 had capsized and one had been lost to enemy action. This count seemed to provoke Fitzroy to make changes to the vessel at his own expense, which improved its seaworthiness.[10] The use of small sample sizes can, however, be misleading. De Moivre's equation relates variability to the sample size and unsatisfactory decisions have been made when possible wide variability associated with small sample sizes is ignored.[11] One approach is to treat these data as 'data' in the singular, which exists as a basis for inference arguments, and not as a general 'fact'. Indeed arguments utilising data can lead to accepted facts.

There is an extensive literature to identify the causes of failures of various types of structure.[12-14] These come to the uniform conclusion that the failure rate is two orders of magnitude greater than the notional probability of failure, p_N, anticipated by the codes of practice. With little evidence that refinements in codes of practice make much difference to the chance of technical failures (G not crossing G* in Fig. 2), it has to be construed that the excess failures must be associated with p_{TU}, p_{UT} and p_U. These predominant causes of failures are lumped together in the literature as failures owing to errors, which are defined as departures from accepted practice as opposed to departures within accepted practice. Fig. 3 shows a Venn diagram which displays the occurrence of errors on the structures in the domain, D. D is divided in two ways: first between failed, F, and safe, S, structures as in Fig. 1, with an unknown boundary between F and S of G*, and second between structures with and without errors, E and R, separated by the boundary P. The intersect of F and E is termed K and represents the failed structures that incorporate errors and that between F and R the error free structures that failed. An indication of the importance of errors in assessing structural failure is that

3	$$p_T \ll p_K \approx p_{FO}$$

A study of the incidence of error induced failures (K) results in some unambiguous statements,[14] namely

(a) errors cannot be eliminated and are, at this time, largely unidentifiable before construction
(b) engineers and contractors are responsible for most errors; those by engineers are attributable to omissions in their professional preparation, experience and mistakes, and by contractors are a result of ignorance, neglect and thoughtlessness
(c) multiple errors are usually needed to produce failure.

An obvious source of errors is in the design calculations. Studies of checking of calculations[15,16] show, however, that this is not a rich harvest. There appears to be little reward in eliminating design errors by the increase of checking time and costs. Little doubt exists that errors involve human behaviour[17] and depend on professional environments. The following lists many disturbing professional practices

(a) inadequate experience of and rewards for design and field personnel, especially when novel structures are considered
(b) communication difficulties between the design–construction–operation communities
(c) absence of scepticism about codes and calculations
(d) frequent personnel changes and compressed design–construction schedules

(e) practice and reality of cutting corners.

These practices show an overlap with the non-technical climates of Pugsley.[1]

It is not, however, apparent that the lay public has a clear understanding of risk in this sense. There is little evidence that the structural engineering profession is any better. The level of acceptance depends very much on the consequences of the failure. If not catastrophic, as is usual in structural failures, then the event will be tolerated more readily than a dramatic one. The combination of failure rate, owing to technical or error causes, and the consequences suggests that risk is the critical measure. The understanding of risk is, however, neither apparent in the public nor in the structural engineering profession. Certainly the public seldom discerns the difference between technical and error caused failures. Indeed there is no reason why it should. As with the professional engineer, it does not anticipate such errors and their occurrence is uniformly underestimated.

The chances of failure caused by technical shortcomings are small and in line with the expectation in design codes. Multiple errors in design, construction and operation provide the main cause of failure and are associated with non-technical environments. The technical environment abounds with the language of uncertainty (probability, statistics) and yet the consequences of technical work appear to be almost perfectly certain compared to the uncertainties of non-technical impacts on structural safety. Most failures are the result of non-technical errors, which are seldom considered by practitioners and require consideration of uncertainty to achieve improved levels of structural safety.

4. UNCERTAINTIES

In Fig. 1 the structural domain is separated into two sets, safe (S) and failed (F) separated by the boundary, G*. This is the world of complete certainty in which structural engineers would like to operate. It has, however, always been realised that uncertainties exist and the situation of Fig. 2 is accepted. Here F and S are uncertain and G is an estimate of the elusive G*. Early codes of practice made design provisions by specifying stresses that, when exceeded, led to failure and when respected resulted in safety. These limiting stresses provided for strength, deflections and stability concerns. The stresses specified had factors of safety built in of between 5 and 6 and F, S and G of Fig. 2 could be confidently treated as the F, S and G* of Fig. 1. Designers therefore proceeded in a certain world in which they somehow felt 'protected' from disasters by safety factors which accounted for, amongst many features, ignorance, extreme loadings and strengths, and social implications.[18]

More recently uncertainties have gained some specific recognition in the design codes. This can be visualised by using the boundary G to estimate G* in Fig. 2. Uncertainties in F, S and G are now considered and notional probabilities of failure described by reliability indices (the inverse of the coefficient of variation of the distribution of G). The intentions of the codes have, however, changed from attempting to include a broad coverage of structural safety to a narrower interpretation that concentrates on p_T alone as expressed by the notional probability of failure.[19] Essentially an attempt to include a complete coverage of structural safety in codes has been replaced by greater precision about the technical aspects.

Most instruction in structural design considers ultimate failure and is based on load and resistance factor design (LRFD) codes.[20] The two approaches to safety evident in earlier and contemporary codes are linked by the calibration of the latter to ensure the same structural performance as the earlier ones. By this calibration the present codes carry the non-technical implications of the predecessors into their target probabilities of technical failure. This puts the designer in the position of taking either approach and obtaining much the same resulting structure. As stated previously, the notional probability of failure approaches p_T, which is a couple of orders of magnitude smaller than p_F. This means that Fig. 2 only reveals a partial picture of the world of uncertainty. Fig. 3 is intended to complete the understanding and all that is omitted previously is explained by the region E on the structural domain, D.

The region E covers errors of commission and omission. p_{TU} in equation (1) involves errors with respect to technical matters. These do occur: the misreading of steel tables and getting the statics wrong are examples of documented failures. p_{UT} covers failures caused by what the technical professional should know and yet does not employ. This form of ignorance, such as the ignoring of the lateral behaviour of suspension bridges, has occurred. Both p_{TU} and p_{UT} will be a continuing part of professional practice. They involve errors of commission and are collectively much smaller than failures due to errors of omission, p_U, which involve non-technical matters that can be significant in structural failures.

Table 1 gives estimates of the time to design a tonne of structural materials in various types of structures. It provides an insight into these errors of omission. The structures of recent origin such as space craft, aircraft and offshore oil drilling structures require much more design time than structures with which there is a long experience. In such modern structures the question 'what can happen here?' as opposed to 'what did happen here?' has to be asked. In traditional structures there is an abundance of past experience to provide guidance. Without that experience there is a necessity to take into account a wider range of concerns than in traditional structures. This involves more time and contributes to the ordering in Table 1. These broader concerns can reduce the uncertainties leading to p_U. In the case of traditional structures broader concerns are generated by experience of surprising events. Examples are the Ronan Point failure and the subsequent concern for progressive collapse and

the Kings Street bridge failure and the concern about higher operational glassy temperatures in weld material.[6]

Of course, many of the troubles with spacecraft have not been beyond the understanding of more traditional structural engineers. Bits falling off and causing structural damage, details not functioning at operational temperatures and supplies, such as a telescope lens, being incorrect come to mind. There is an indication in Table 1 that considerations in novel structures will involve much more time with respect to the reliability associated with p_U than in traditional structures and that omission errors will be reduced. Certainly in professional situations where the avoidance of hazards is critical, such as in the nuclear power industry, the question 'what can happen here' dominates the discussion and the associated design time is high compared to that expected in buildings and bridges.

The uncertainty of Fig. 3 can be viewed in three ways, aleatoric (arising from statistical variation), epistemic (arising from modelling limitations) and ontological (to do with the difference between an engineer's assumptions and reality).[21] Aleatoric uncertainty is about countable information available. This is often treated as random and arranged statistically. This information can be treated as data which are employed as the basis of inference and the establishment of models which allow significant generalisations beyond the a priori data to be generated. This modelling of data, which can enhance understanding of structural safety, composes epistemic uncertainty. The completeness of the data on which the inference is based is of critical importance and the domain of concern must be understood. Aleatoric and epistemic uncertainty are the main ingredients of contemporary codes of practice.

Epistemic uncertainty considerations are successful in assessing the technical part of failure, p_T, in equation (1) and aleatoric uncertainty data provide an estimate of the combined p_{TU} and p_{UT} contributions to failure. In equation (2), however, it can be observed that failures are largely caused by factors that are not considered in structural safety. These are omissions, that is factors which are not considered in structural safety, and commissions, that is actions that are in error and may range from simple mistakes to malicious acts. These errors of omission and commission fall under the umbrella of ontological uncertainty. It is here, as expressed by p_U, that most failures originate and where little design time of conventional structures such as bridges and buildings is employed. Indeed, as pointed out in section 2, it is not evident where authority for such considerations lie.

The separations of uncertainty indicate that the reduction of p_U ensures the greatest increase in structural reliability. Table 1 shows the great design time put into spacecraft compared to other situations. One reason was the necessity of including all aspects that could reduce structural safety. This had to be done without past experience and the studies were dominated by problems of ontological uncertainty. Essentially the question was 'what can happen here?'. The same happened with aircraft in the first half of the 20th century. This changed with the introduction of the Messerschmitt BF 109 in 1935 which, with its monocoque metal body, retractable undercarriage and

Structure	Design time: h
Spacecraft	4.5×10^6
Aircraft	9×10^4
Offshore structures	
Novel	90
Conventional	9
Bridges	6
Buidings	2

Table 1. Design time for each tonne of structure

enclosed canopy, set the pattern for subsequent fighter planes. Over 30 000 of these were built.[22] From then on experience became available and the problems of such aircraft became codified. This codification occurred in traditional structures (buildings and bridges) at the beginning of that century.

With ongoing experience most eventualities could be provided for in revised codes of practice. The ontological uncertainty was only explored as new failure situations were observed. Examples are the impact of aircraft on structures, the drift of high rise buildings when masonry cladding was replaced by light weight cladding[23] and vulnerable details after a significant earthquake. The question asked was 'what did happen here?' and only if the answer indicated novel safety concerns was there a move to broaden the studies. It would seem that in such structures acceptable levels of safety or economics have been achieved and there is little social and professional impetus to explore ontological uncertainty. One reason for an awakening interest may occur because of the success in reducing the death toll owing to structural failure in the technically advanced world. Deaths resulting from earthquakes and sea flooding have been reduced by orders of magnitude and the present concerns are about economic damage caused by structural failures. These concerns may stimulate a new effort to include ontological studies.[7,24]

In assessing and achieving structural safety both surprise and uncertainty about the results have to be minimised. They are not the same measures but it is only in the realm of uncertainty that surprise will flourish. This is not the same as searching for low probability environments. Although surprise has a low probability of occurring, not all low probability situations are surprising.[25] An example is the absence of surprise, as opposed to gloom or pleasure, in any hand dealt in the game of bridge. The probability can be the same as a coin landing on its edge when tossed; an event which is surprising. It is sensible that studies on improving safety assessment concentrate on areas which have not been exhaustively considered, and may uncover surprises, and avoid focusing on issues which at best could deliver only very minor improvements. Inasmuch as p_U has much the same value as p_F, it would be anticipated that most studies would concentrate on ontological uncertainty. Any reading of the literature shows, however, an emphasis in reducing the distance between G* and G in Fig. 2. Elms[26,27] has described a procedure to obtain consistent crudeness in understanding the various uncertainties evident in assessing structural safety. Any application of this theory of consistent crudeness can but justify the enthusiasm of Blockley to look 'outside the box.'[6,7] Indeed the broader climates espoused by Pugsley over 30 years ago[1] should still be dominant today.

5. DECISIONS

Professions have a common activity of making decisions in the face of uncertainty in a finite, limited period of time. The impact of the uncertainty and the time constraint make it likely that errors will occur. Structural engineering is no exception to this situation. Table 1 provides a sense of the relative time spent on designing different structures. In the cases of spacecraft and aircraft the design objectives can be different from the other structures listed. Both of these have design requirements of speed and acceleration as well as the pay load.

The power and fuel needed will depend on these and the vehicle weight. Newton's second law, $F = ma$, is the basis for the design decision which can be termed a minimum weight optimisation scheme. In this way the mass, and hence the power requirement, F, is minimised.

The other structural types listed in Table 1 are based on $F = 0$ and optimisation with respect to weight has no attractions. There is a literature that advocates optimisation with respect to life time costs for such structures[16] which demands the specification of the probability of failure, discount and interest rates, future costs and benefits, costs of failure involving legal obligations when there is a hint of error and conversions from money to utilities. Except for p_F these estimates of the future are influenced, if not dominated, by the musings of economists. Cost-benefit decision makers are often accused of justifying any previously decided course of action. This may be based on personal opinion introduced into these subjective specifications that model the future. If this is to be the course of decision making about structures then the work of Elms on consistent crudeness,[26,27] whereby attention is directed to the features in the optimisation scheme that are least understood, is of importance. To the extent that optimisation with respect to costs and benefits occurs it is at levels of responsibility at which the structure is only one influential element. Structural failures can affect the operation but they are not the dominant concern of the decision maker. Worries about structural failures fall within the province of structural and geotechnical engineers where there exists a literature on formal analytical schemes[20,28] that ensure consistency and order.

Risk involves a combination of the chances of a failure and the consequences. For precision the various economic forecasts in a cost-benefit scheme are required but, for a comparative study between locations and events, consistency can be more important. Thus the consequences of an earthquake in New Zealand may be considered ominous compared to the same earthquake in similar places in Australia or the USA. Among other matters, the impact of the event on the national economy and the critical nature of the location in that economy have to be included. These matters can often be assessed in a defendable manner and the balancing of risk between locations can be attained by changes in frequencies of occurrence in codes of practice and other institutional guides to professional practice. Consequential evidence may indicate preventative and also precautionary measures that reduce the chance of failures. For instance, retrofitting against earthquakes and the terminating of fatigue cracks by holes are preventative actions that reduce the chances of failure and use of sensing devices on structures is a precautionary action that give warning of structural changes. These are costly strategies which structural engineers may employ when justified by dire consequences of failure.

There appears to be a procedure by which professional decisions are made. This starts with a basic understanding of a broad field in which the topic (structure, legal case, patient) is embedded In a Bayesian model; this would be the *a priori* understanding. The particulars of the topic are then introduced and are analysed and argued about in the light of the earlier understanding. A decision is made in a finite time on the combined picture of the general and particular. Uncertainty exists in all these professional situations and the ontological part is of great importance. There is never a complete coverage of variables and possibilities. If errors are to be made in the decisions, they largely occur because of these omissions. An example is the identification of culprits in legal cases which subsequent DNA analysis showed to have been wrong.

Uncertainties have much in common from profession to profession. A medical internist and a geotechnical engineer examine the surface and obtain internal samples by biopsy or boring. These provide local internal evidence which have uncertainties of the aleatoric and ontological types. Decisions are made in the light of these uncertainties and the probability of occurrence enters the discussion. Legal trials in the English–American system empower a jury to make decisions based on levels of probability that are different for civil and criminal cases. Here it is likely the understanding of probability is not that of axiomatic theory but is in terms of personal interpretation. It can be expressed in lay terms that make sense in a subjective framework. It is considered that this also has virtue in structural decision-making just as it has in legal situations. This power, in spite of shortcomings, has persisted for about a thousand years and does not seem likely to be supplanted. This suggests that it enjoys public and legal confidence even if the understanding of probability is a personal one. In much the same manner probability is not always used with precision in structural decision making.

Again it is likely that the understanding is personal and can be expressed in lay terms that make realistic sense. It has virtue in structural decision making as in legal ones. In this respect Cohen[29] has argued that a Baconian view of probability in which the completeness of the evidence is graded, is applicable, as opposed to the Pascalian view where the evidence is complete. This separation of probabilities has a parallel to 'open' and 'closed' views of systems analysis.[30] The open system includes matters that are outside the immediate environment of the considered problem and yet which might influence the outcomes in some, often undefined, manner. The closed system is amenable to the axiomatic constraints of probability theory but requires a complete specification of alternatives. With openness an inclusion of matters that have to be considered in assessing UT and U failures is possible but the comfort of precise assessment values is lost. With a closed situation the problem is well defined and the results sharp. It is well suited to avoiding T type failures. Both approaches to probability can have a place in structural reliability.

Decisions are made in structural design in two stages. The responsible engineer decides about the structural system. In the case of buildings the architect usually has defined what is possible and the engineer has less control than in bridges. The relative control in decision making may explain the design time shown in Table 1. Additionally the value of the structure in a building (about 15% in a high rise one) is smaller than in a bridge. The decision by the responsible engineer is an iterative process which can be likened to painting a picture. The start is with a blank sheet of paper and the final scheme is determined by trial and error, an understanding of alternatives, materials and objectives, and a synthetic ability. In the end a decision on the final scheme is made and the work of proportioning members and connections commenced. This is the second design activity and today is governed by codes of practice and in the earlier days by a protocol established by the design firm.

For much of the design both by the responsible engineer and by the design staff, the process can be thought of as satisficing rather than optimising. Formal decision making is organised into the establishment of alternatives that satisfy objectives and constraints, and criteria that allow the selection of the best alternative. The claim of satisficers[31,32] is that the human mind is incapable of modelling all significant practical problems; the alternatives, objectives and constraints are incomplete and a formal decision process cannot produce global answers. To solve real-world problems, solutions that are acceptable to the stakeholders are advised. The seeking of acceptable as opposed to best solutions is termed satisficing. In architecture it has been claimed that the best facility is the one with the least problems or conflicts.[33] The uncertainties previously discussed indicate that there are good reasons why satisficing is what much of structural design, like architecture, is about.[34] This means that the search for a design that satisfies various wishes such as safety, resilience, durability and an insensitivity to changes of variables and parameters, can better describe structural decision making than one that seeks to optimise. Indeed, the establishment of a satisficing environment is evident in codes. Table 2 shows the annual frequency of failure implications of Australian and New Zealand codes for wind, earthquake and snow loading.[35] At first sight it seems strange that there is no consistency in notional probability of failure either between the loadings or, for earthquake and snow loadings, between countries. The codes are not balanced either internally or between each other. The values of p_T are inconsistent. The issue is significant and warrants further discussion.

To begin with, all failures are not equal. Failures owing to wind, earthquake and snow have different characteristics and different economic and community implications. A recounting of some of these differences in engineered structures is instructive.

(a) Earthquakes produce total structural failures, wind and snow predominantly produce local failures of roofs.
(b) Earthquakes strike without warning whereas heavy snow loads and high winds are more predictable.
(c) Damage caused by earthquakes is likely to be more severe with respect to the local infrastructure than due to wind and snow. Serious damage and disruption to water supply

Design life: years	Annual frequency of exceedence		
	Wind	Earthquake	Snow
25	1/1000 *1/1000*	1/250 *1/1000*	1/1000 *1/250*
50	1/2500 *1/2500*	1/500 *1/2500*	1/2500 *1/500*

Table 2. Intended annual frequency of exceedence of Australian and New Zealand (*italics*) code provisions

and transportation is demonstrably more likely caused by earthquakes than by wind and snow.

(*d*) The time to recover from earthquakes and wind is longer than from snow storms. As a result of these economic differences and the extent and duration of the three disasters, the subsequent functioning of the communities will be different.

(*e*) The threat to life is high in earthquakes, moderate in winds and low in snow storms.

In reaching their decisions, code committees have to bear these issues in mind. Besides these differences in the nature and consequences of failures, there are other issues including

(*a*) the need to calibrate new codes to previous successful and accepted practice, which, in turn, could be seen as an encapsulation of experience

(*b*) response to previous failures, especially recent ones, and the recommendations of commissions of inquiry

(*c*) pressures from industry and professional groups

(*d*) the tendency of regulators to make regulations more stringent than previous ones and of code writers to make codes even less stringent, both without compelling reasons

(*e*) a general and appropriate desire to feel comfortable about the results.

This Pugsleyan listing makes it scarcely surprising that code requirements display apparent inconsistencies. Clearly the design process is one of satisficing rather than optimisation. Two questions do, however remain

(*a*) could the codes be better?

(*b*) should their values be consistent?

A second issue arising from the apparent inconsistencies in Table 2 is the relationship between p_N and p_T. Elsewhere it has been stated that the probabilities converge and tend towards common values. This may not be for a teleological reason. The notional probability is a mental construct whereas the probability of technical failure is a real, if elusive occurrence. p_N is determined for reasons that are ill defined and often unstated such as the reduction of surprise and avoidance of failure when errors exist. On the other hand, p_T is difficult to separate from the three other failure probabilities and to all intents and purposes impossible to measure in practice.

A reason for this convergence could be that p_N tacitly defines the average p_T for a class of structures and within this class the range of p_T depends on the level of refinement of the code. Thus as codes 'improve' the best estimate of G*, it approaches closer to G in Fig. 2.

6. DISCUSSION

Over the life of a structure decisions are made that may involve planning, design, bidding and acceptance, construction, acceptance, operation, maintenance, rehabilitation and changes and removal. As these stages progress the chances of failure alter.[8] Until actual construction is underway failure is in the mind rather than a reality. Over the rest of the life failure can, however, occur; the chance of failure changes with time and depends on the ownership. The owner of a highway system or a retail chain has a different responsibility for life use of a structure than a developer. In the first situation society is

somewhat protected against failure by the continuing interests of the owner, whereas, in the second situation the protection may be shared with local public authorities. In either situation continual vigilance and assessment are essential ingredients in achieving structural reliability.

Failures can be divided into two classes: one confined to an individual structure and one, such as an earthquake or act of war, that affects many structures. In the first case the ideas of the previous paragraph prevail and although the loss may have a wide impact, the ownership is specific and isolated; the remedy is local. In the second case the losses and ownership are wider. The responsibilities are different in the two situations and the domain, D, has to be viewed differently. The professional efforts with respect to the individual structure have been dominated by the saving of lives. This is certainly of interest when considering the many structures affected by the wrath of nature and war, but additionally the broader impact on the economics and social arrangements have to be included. Here structures which have been damaged are often as important as those that have collapsed. The remedies are seldom local and demand the resources of a much wider community, usually by means of taxes and levies. This intrusion of politics and other topics of the social sciences ensures that the climates of Pugsley[1] are explicitly important in the achieving of structural safety. The existing arrangements for attaining safety over the lifetime of individual structures appear to be satisfactory to the profession and to society as whole. The failure rates and consequent loss of lives do not appear out of line with the normal dangers of everyday life and do not inspire public outcries. Professional endeavours to improve this form of structural safety tend to concentrate on technical matters to reduce p_T and to shift G closer to G* in Fig. 3. There is no doubt that this professional work has been successful.

Concerns about the safety of individual structures have usually been due to surprising phenomena and the increase of p_U. Once displayed such a new phenomenon receives intense, and often over enthusiastic, professional attention and acceptable levels of safety quickly result. Where many structures are involved there is a shorter history of professional experience. Riley[3] has ordered the inclusion of all stakeholders into the discussions of future civil work of the US Corps of Engineers. In the case of the effects of earthquakes and flooding this would include medical and transportation professionals as well as those providing power, fire protection, rescue and other services. There is certainly a place for structural engineers in these deliberations and from the discussion of Table 2 it is clear that they have responded to these widespread events by the local requirements for safety.

It is not, however, clear how structural engineers will be placed in the integrated planning and decision making. It is certainly a time for the professional institutes to take a leadership role. This should at least include an extensive non-technical interplay with professional institutions whose membership have a stake in the decisions and, more likely, the subjection of individual institutional authority to broader communities such as the national academies.

A virile profession requires a continual environment of change.

This is the case with structural engineering. For instance, changes in the construction, the continual assessment of structural performance and the monitoring of structural behaviour are documented.[36-38] There are also changes in professional environment which may be subtle but effective. One is the impact of commercialism into professional life where pressures, other than those traditionally accepted in practice, appear in the workplace. These are novel situations and Pugsley's list is particularly applicable. The institutions have to be unambiguous in their concerns about the effects of changes in practice in achieving structural reliability. One source of information is the informed and valid whistleblower. When such accounts are proved correct, the professional institutions should protect the informants. This could mean a change from past policies.[39]

The place of errors in assessing and achieving structural safety, as indicated by equation (2), has been significant. An acceptance of human failings that lead to errors exists in the population.[40] This generosity is more evident when the consequences are not serious. Fortunately most structural failures are local with limited consequences and public blame is seldom displayed. Optimism is the order of the day with respect to most human endeavours and structural engineering is no exception.

An essential aspect in achieving structural safety is the combination of quantitative and qualitative insights. Something can be learned from the risk assessment methods conventionally practiced in hazardous industries. Before considering chances or consequences the question 'what can or could go wrong?' is addressed. The answer probes into past experience and is an attempt to discern reality. Professionals from all relevant disciplines are at the table. In the case of a bridge these would at least include staff functioning in design, construction, maintenance and operation. This is a stage when experience is truly valued and where 'forgotten' events can be revealed.[40] The subsequent decisions will rest heavily on this preamble.

An example of the importance of ascertaining real world scenarios is cited by Stephens[41] when concerned with the unintended launching of nuclear tipped missiles. The considerable confidence obtained in initially estimating on the basis of paper constructs the safety of the system was squashed when the real evidence of experience was included. The procedure of determining what was in reality known before proceeding with calculations provided confidence in predicting behaviour. This separation of design into these two parts, the search for what is known and the actual production based on contemporary codes, can be related to Fig. 2. The search for reality is an attempt to identify locations near or on G*. The actual production is an attempt to provide a structure located inside G. The first part can help ensure that available evidence from the past is not overlooked and that UT type failures do not occur.

Basically, to deal with the non-technical sources of error by structural engineers requires a different outlook and a different training. Perhaps the tasks and roles of the structural engineer need to be revisited, and here Blockley may be suggesting a way with his advocacy of 'doing it differently'.[42]

Encouragement is found in the work of the Standing Committee on Structural Safety (SOCC)[43] and the emphasis on 'soft management' issues. This view is espoused from a business and industry perspective, which could augment the professional concern of this paper and the discussions and rejoinder to Melchers' study (Ref. 8).[44]

The independence between T, TU, UT and U is important in this discussion on structural safety. These definitions of the causes for failure were introduced by Melchers et al.[45] and their relationship, as displayed in equation (1), in Ref. 8. The use here emphasises the importance of ensuring that the considerations of errors has been complete. Comparison of this work with ideas in another field at another time is of interest. In 1689 John Locke[46] stated the roots of human error as

(a) want of proof
(b) want of ability
(c) want of will
(d) wrong measures of probability.

In this paper the roots of errors of commission and omission are

(a) TU, of a technical nature
(b) UT, associated with technical ignorance
(c) U, caused by events outside the professional environment.

The first two in each of these lists can be construed as equivalent. The want of will is apparent in the efforts needed to avoid technical errors, overcome ignorance and to be involved in matters beyond the technical realm. Over 200 years ago probability measures were tied to the likelihood of a fact or idea being true in the light of evidence and did not demand completeness. They were open systems. Today the meaning is applied to a closed system and supported by the axiomatic definition of mathematics. Two types of error can be identified from Locke's last root, omission of evidence and exclusion of alternative facts and ideas. These were certainly the overwhelming concerns with respect to U type errors discussed here. It is claimed that the broad view of this paper covers the various origins of errors.

7. CONCLUSIONS

Technical matters are the province of structural engineers. Codes are relied on for the avoidance of technical failures, T. A quality assurance process must, however, be in place to ensure that the codes are followed. This is the responsibility of civic and state authorities as well as an internal imperative for an engineering firm. The aim is to ensure that in effect technical failures are virtually impossible.

Errors dominate the ability to assess and achieve structural safety. Three categories are evident.

(a) TU, technical errors; again a checking process must be in place and applied to even minor structural parts. Failure of minor parts can have great financial consequences; here the Hyatt Regency Hotel failure and the Mississippi bridge in Minneapolis collapse are prime examples. These are matters for the professional engineer and can be reduced by changing managerial arrangements and require additional professional resources.

(b) UT, errors of ignorance about technical matters; these require the individual continuing study and the time and opportunity for such study if this personal professional ignorance is to be reduced. The responsibility here is with the individual, the professional management and the professional institutes. Errors of this type can apply to the profession as a whole, for example, when shortcomings in codes of practice are suddenly revealed by previously undisplayed failure modes. Codes are quickly modified to remedy this error. The professionals involved in code writing have responsibilities not only to respond to these sudden events but also to anticipate them. This may require examination of broader issues at the expense of attaining greater precision. Certainly when new or different ventures are considered the climates of Pugsley[1] must be incorporated. This requires the professional environment and training to be much extended.

(c) U, errors of ignorance of the impact of non-technical matters on structural safety. The main professional thrust has to be the avoidance of surprise. Most failures can be assigned to this source and yet there are seldom existing organisations to provide the necessary extra-mural insights. To reduce surprise requires the leadership of professional institutes which must be actively involved with non-engineering communities. Such involvement will provide insight to others on the role of engineers and a professional understanding of the external impacts on structural safety. The dissemination of this understanding is in the lap of professional institutes.

The relationship of these errors to the assignment of responsibilities, modes of failure, uncertainty and the making of decisions comprises the body of the work. To broaden the consideration of safety it will be necessary to enlarge the understanding of professional engineers so that they are aware of the ideas and vocabularies of contributing disciplines.

REFERENCES

1. PUGSLEY A. C. The prediction of proneness to structural accidents. *Structural Engineering,* 1973, **51**, No. 6, 195–196.
2. BROWN C. B., JOHNSON J. L. and LOFTUS J. J. Subjective seismic safety assessments. *Journal of Structural Engineering, American Society of Cvil Engineers,* 1983, **110**, No. 9, 2212–2233.
3. RILEY D. *Shared Vision Planning.* Memo to District and Battalion Commanders, United States Corps of Engineers, 2007.
4. CONRAD J. (ed.) *Society, Technology and Risk Assessment.* Academic Press, London, 1980.
5. ANDREWS C. E. *Final Report on the Tacoma Narrows Bridge.* Report to the Washington Toll Bridge Authority, Olympia, Washington, USA, 1952.
6. BLOCKLEY D. I. *The Nature of Structural Design and Safety.* Ellis Horwood, Chichester, 1980.
7. HENDERSON J. R. and BLOCKLEY D. I. Structural failures and the growth of engineering knowledge. *Proceedings of the Institution of Civil Engineers, Part 1,* 1980, **68** (November), 719–728.
8. MELCHERS R. E. Structural reliability theory in the context of structural safety. *Civil Engineering and Environmental Systems,* 2007, **24**, No. 1, 53–69. (See also Discussion and rejoinder in Ref. 44).
9. SCHWEBER L. *Disciplining Statistics. Demography and Vital Statistics in France and England 1830–1885.* Duke University Press, Durham and London, 2006
10. BRENT P. L. *Charles Darwin: A Man of Enlarged Curiosity.* W. W. Norton, New York. 1981.
11. WAINER H. The most dangerous equation. *American Scientist.* 2007, **95**, No. 5, 249–256.
12. MELCHERS R. E. Structural reliability assessment and human error. *Reliability and Risk Analysis in Civil Engineering. Proceedings, ICASP-5.* Institute of Risk Research, Vancouver, 1987, Vol. 1, pp. 46–54.
13. NOVAK A. S. (ed.) Modeling human error in structural design and construction. *Proceedings of a National Science Foundation Workshop,* Ann Arbor, 1986.
14. BROWN C. B. and YIN X. Errors in structural engineering. *Journal of Structural Engineering, American Society of Civil Engineers.* 1988, **114**, No. 11, 2574–2593.
15. MELCHERS R. E., HARRINGTON M. and STEWART M. G. Human error in structural reliability. *Civil Engineering Reports 1–V.* Monash University, Australia 1983–1985.
16. MELCHERS R. E. The influences of control processes in structural engineering. *Proceedings of the Institution of Civil Engineers,* 1978, **65**, No. 2, 791–807.
17. Reese R. C. Structural failures from the human side. *Civil Engineering, American Society of Civil Engineers,* 1973, **43**, No. 1, 81–84.
18. BROWN C. B. Concepts of structural safety. *Journal of the Structural Division, Proceedings of the American Society of Civil Engineers,* 1960, **86**, No. ST-12, 39–57.
19. ELISHSKOFF I. *Safety Factors and Reliability: Friends or Foes?* Kluwer Scientific Publications, Dordrecht, 2004.
20. MELCHERS R. E. *Structural Reliability, Analysis and Prediction.* Ellis Horwood, Chichester, 1987.
21. ELMS D. G. Structural safety: issues and progress. *Progress in Structural Engineering and Materials,* 2004, **6**, No. 2, 116–126.
22. GLANCEY J. *Spitfire. The Biography.* Atlantic Books, London, 2006.
23. WOOD R. H. The stability of tall buildings. *Proceedings of the Institution of Civil Engineers,* 1958, **11**, 69–102.
24. BROWN C. B. and ELMS D. G. Structural design codes of practice and their limits. *International Journal of Risk Assessment and Management,* 2007, **7**, No. 6/7, 773–786.
25. WEAVER W. *Lady Luck. The Theory of Probability.* Anchor Books, Doubleday & Company, Garden City, New York, 1963.
26. ELMS D. G. The principle of consistent crudeness. *Proceedings of the NSF Workshop on Civil Engineering of Fuzzy Sets.* Purdue University, 1985, pp. 35–44.
27. ELMS D. G. Consistent crudeness in system construction. *Optimisation and Artificial Intelligence in Civil Engineering* (TOPPING B. H. V. (ed.)). Kluwer Scientific Publications, Dordrecht, 1992, Vol. 1, pp. 61–70.
28. JORDAAN I. J. *Decisions under Uncertainty. Probabilistic Analysis for Engineering Decisions.* Cambridge University Press, Cambridge, 2005.
29. COHEN L. J. *The Probable and the Provable.* Clarendon Press, Oxford, 1977.

30. STEWART M. G. and MELCHERS R. E. *Probabilistic Risk Assessment of Engineering Systems.* Chapman & Hall, London, 1997.

31. SIMON H. A. *Models of Man, Social and Rational: Mathematical Essays on Rational Human Behavior in Social Settings.* Wiley, New York, 1957.

32. SIMON H. A. *Administrative Behavior,* 2nd edn. Free Press, New York, 1957.

33. ALEXANDER C. J. *Notes on the Synthesis of Form.* Harvard University Press, Cambridge, 1964.

34. Brown C. B. Optimizing and satisficing. *Structural Safety,* 1990, **7**, 155–163.

35. REID S. G. Design differentiation for critical infrastructure networks for disaster mitigation. *IFED07, Shoal Bay, Australia,* 2007.

36. DESTER W. S. and BLOCKLEY D. I. Managing the uncertainty of unknown risks. *Civil Engineering and Environmental Systems,* 2003, **20**, No. 2, 83–103.

37. STEWART M. G., ROSOWSKY D. V. and VAL D. V. Reliability-based bridge assessment using risk ranking decision analysis. *Structural Safety,* 2001, **23**, No. 4, 397–405.

38. DAS P. C. Development of a comprehensive structures management methodology for the highways agency. *Management of Highway Structures.* Thomas Telford, London, 1999, pp. 49–60.

39. BROWN C. B. Trapped in one's own times: the individual's relationship to education and practice. *Civil Engineering and Environmental Systems,* 1996, **14**, No. 1, 1–14.

40. SILVERMAN D. *The Theory of Organizations.* Heinemann, London, 1974.

41. STEPHENS K. Using risk methodology to avoid failures. In *Owning the Future* (ELMS D. G. (ed.)). Christchurch, New Zealand, Centre for Advanced Engineering, 1998, pp. 303–308.

42. BLOCKLEY D. I. and GODFREY P. *Doing it Differently: Systems for Rethinking Construction.* Thomas Telford, London, 2000.

43. STANDING COMMITTEE ON STRUCTURAL SAFETY. *Executive Synopsis, 16th Annual Report.* Scoss, 2007, pp. 1–11. See www.scoss.org.uk/publictions.asp for further details.

44. BLOCKLEY D. I., ELMS D. G., BROWN C. B and MELCHERS R. E. Discussion and closure to Ref. 8. *Civil Engineering and Environmental Systems,* 2008, **25**, No. 1, 59–76

45. MELCHERS R. E., BAKER M. J. and MOSES. F. Evaluation of experience. *Quality Assurance within the Building Industry.* International Association of Bridge and Structural Engineering, Zurich, 1983, pp. 21–38. Report No. 47.

46. LOCKE J. *An Essay Concerning Human Understanding* (NIDDITCH P. H. (ed.)). Oxford University Press, New York, 1979.

What do you think?

To comment on this paper, please email up to 500 words to the editor at journals@ice.org.uk

Proceedings journals rely entirely on contributions sent in by civil engineers and related professionals, academics and students. Papers should be 2000–5000 words long, with adequate illustrations and references. Please visit www.thomastelford.com/journals for author guidelines and further details.

Proceedings of the Institution of
Civil Engineers
Structures & Buildings 161
August 2008 Issue SB4
Pages 231–237
doi: 10.1680/stbu.2008.161.4.231

Paper 700036
Received 10/07/2007
Accepted 13/02/2008

Keywords: management/quality
control/risk and probability analysis

David Blockley
Senior Research Fellow,
University of Bristol, UK

Managing risks to structures

D. Blockley PhD, DSc, FREng, FICE, FIStructE

Structural engineering has changed markedly over the last decades, creating new challenges and new opportunities. Consequently, structural engineers are widening their thinking from just technical issues to the effects of other matters on the risks to their structures. In particular there is a need to find better methods for integrating hard and soft risks. Hard systems are physical and technical matters traditionally dealt with by engineering science. Soft systems involve people and include matters traditionally dealt with by engineering management. In order to make improvements, engineers have to combine good-quality evidence from disparate sources, both technical and from wider issues. The current paper demonstrates how disparate evidence can be measured and combined using interval probabilities drawn as colourful 'Italian flag' indicators of risk. Process models are used to map the progress of projects. An Italian flag is associated with each process to indicate the level of dependability, based on all the information available at the time, that the process will be successful, which is to reach the stated objectives. A new method of pairwise combinations is described and used to calculate the flags through the entire process model. An example of the procurement of a building is used to illustrate the method.

1. INTRODUCTION

Structural engineering has changed markedly over the last decades. Not only are new powerful computational tools available, but there are different challenges such as terrorist attack and the impact of climate change. Consequently there is a much greater awareness of the need to deal explicitly with risk in all aspects of life. A key difficulty is how structural engineer decision makers integrate information from many disparate sources to manage the risks to structures; this is the central topic of this paper.

2. INTEGRATING HARD PHYSICAL RISKS WITH SOFT 'PEOPLE' RISKS

In many projects, accountants manage the known financial risks well, the engineers manage the known technological risks well, the safety specialists manage the known health and safety risks well, the quality managers manage the known processes well and so on. Major problems, however, even in successful companies, seem to arise in the gaps between these specialisms, resulting in unknown and unintended complications such as

cost and time overruns and consequent quality problems. Even within specialisms, however, a range of techniques may be used to assess different aspects of risk which are difficult to integrate. The problem to be addressed in this paper is how to facilitate good balanced decisions to manage all of the risks and hence minimise unintended harmful consequences: in short to improve the ability to do 'joined up' thinking across a fragmented industry.

Hard systems are physical systems that are commonly said to be 'objective' in that they are supposed to be independent of the observer and hence the same for all of us. Hard systems are the topic of traditional engineering science. Soft systems are, as the name implies, systems that are difficult to define— the edges are unclear. Generally soft systems are governed by the behaviour of people, which can be complex. Soft systems are the topic of traditional engineering management. The emphasis in soft systems is not therefore on prediction, but rather on managing a process to achieve desired outcomes.

Processes are the way things behave in hard systems and what people do in soft systems. All designed hard systems have a function. For example a beam in a structure has the function of carrying the loads from the floor slab. A dam has the function of holding back the reservoir water. The steel and concrete of which the beam and the dam are made does not 'know' it has that function—it has no intentionality. The function is ascribed to a hard system by the people who own it, conceive it, design it, build it and use it. This function can therefore be perceived as a role in a process. Part of defining that role is to decide the criteria of failure. This is done by different people from different points of view and may be contentious. Clearly some functions are obvious; others are less clear and unintended. For example a bridge designed to carry road traffic was almost certainly not designed to be used as a shelter by homeless people. In one case the cost of repair to concrete damaged by the fires lit by homeless people to keep warm under a bridge was substantial.

In summary all designed hard system processes are embedded in one or more soft system processes.

3. PROCESS MODELS

A process model is used as the integrating structure on which everything else is built.[1] There are many views of what constitutes a process. Most people use the word to distinguish the form of the 'doing of an activity' from the content which is the output, for example a product. Here a new way of thinking about a process is applied, which does not separate form and content. This new concept of what constitutes a process is much richer than an input transformed into an output, a recipe, a Gantt Chart, a network or a flow chart, although it can be simplified down to each of these if required. In this methodology each process is seen as a holon, that is it is both a whole and a part at one and the same time. It also has emergent properties,[1] from systems theory, that arise from the complex interaction of the parts. The relationships between processes are captured in a process map and then each individual process is used as a 'peg' on which to attach all other data and information. The process map therefore sets out the basic structure of a project and all of the data associated with it. The traditional separation of process from product is lost because, by this view, a process can be the 'doing of an activity' or the 'performing of a physical system'—the structuring of the information is the same for both.

In order to use this idea the map of the interconnections has to show how processes relate to each other. This is the purpose of a 'project progress map' (PPM): it enables various data such as risk registers, structural calculations, project progress measures and so on to be integrated. As stated above, the processes in the PPM form a central spine or skeletal structure on which all data and attributes are attached.

A product is the output of a process. It is useful to keep these two ideas of product and process quite distinct because it is useful in defining what the clients perceive they are buying. It is important to understand that, because products do things and exist through time, products are also processes seen from this wider perspective.

The process holons are arranged hierarchically in layers. Initially the whole system is described as one process: the 'top' process and the top layer. Obviously to achieve success of this top process many sub-processes have to be successful and these are the second layer. For the sub-processes to be successful some sub-sub-processes must be successful and this continues down to a level of detail as appropriate for the particular system. It is important to realise that this is not reductionism because the many interactions between processes at the same layer are recognised and modelled as far as they are understood. For example time relationships can be included using a standard critical path network analysis.

Each process is monitored to keep it on a path to success. Success is defined by a set of precisely stated objectives. Every process should have a process owner, that is a person who is responsible for delivering success of that particular process. Parameters and performance indicators of the process are monitored to keep them within required bounds and to provide evidence of trends that need remedial action. In this way unintended consequences can be identified early. Thus at any given time during a process evidence is available from past and present performance and predictive analyses about the future can be used. The words 'who, what, where, when, why and how' are formally used to capture and record information for appropriate sharing with authorised other people. In this way each and every process is steered to success based on up to date integrated information.

4. MEASURING EVIDENCE OF PERFORMANCE USING AN ITALIAN FLAG

A key idea of the methodology is to find evidence that any given process is moving towards success and there is no build up of difficulties that might bring about failure. Opportunities for improvement should also be located. The evidence will come from many sources of various types and there is a need to collect it, digest it, interpret it, learn from it and make decisions using it.

In order to have a measure of evidence that can be used in soft as well as hard problems it is necessary to assess, quite separately, the evidence in favour and the evidence against the proposition that a process is heading for success. This can be conceived of in several ways. One way is as a vote by a group. Another is as an individual responsible judgement—a kind of internal vote. A scale of [0, 1] is used as a measure of evidence in favour and evidence against. The degree of evidence that a process is heading for success is coloured in green, as shown in Fig 1.

Evidence against is also assessed on a scale [0, 1] and is coloured in red starting from 1 and working back to zero. The difference in the middle is white and represents incompleteness—the extent to which we do not know. The three colours together make the Italian flag.

An all-green flag means that there is complete evidence for and no evidence against (no red). An all-red flag means that there is complete evidence against and no evidence for (no green). An all-white flag means there is no green evidence for and no red evidence against and so we really 'do not know' or indeed have no view.

Figure 2 shows a simple model of a process with two sub-process holons, the successes of which are together necessary and sufficient for the success of the top process. The players who own each process associate an Italian flag with that process. The flag represents their view, based on evidence, that the process will be successful. Clearly there is something wrong in Fig. 2 since the flags are inconsistent in a rather blatant way for the purpose of the example. This means that when the process owners realise this inconsistency, they can discuss the reasons for it and decide on adjustments or on what needs to be done to improve the chances of success.

Until recently the flags were used entirely qualitatively as described in Ref. 1. Now a new algorithm developed for the purpose of propagating the flags mathematically is described. Certain constraints and assumptions need to be appreciated, however, before the method is used. The new algorithm is implemented in two 'sister' pieces of software called PERIMETA

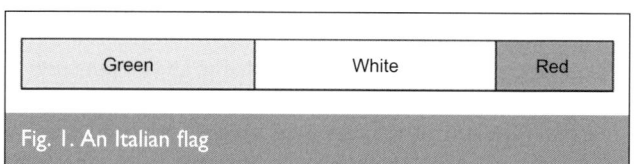

Green	White	Red

Fig. 1. An Italian flag

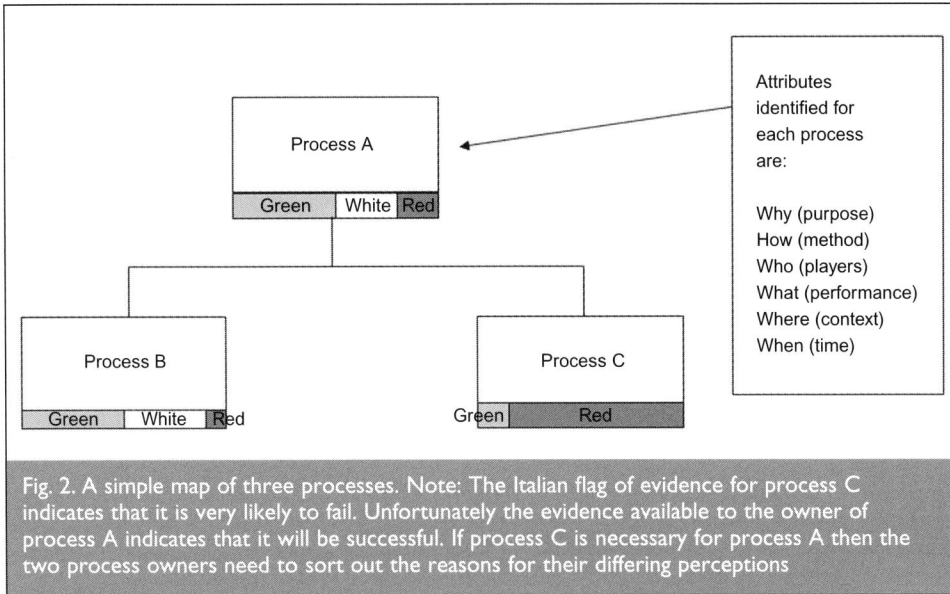

Fig. 2. A simple map of three processes. Note: The Italian flag of evidence for process C indicates that it is very likely to fail. Unfortunately the evidence available to the owner of process A indicates that it will be successful. If process C is necessary for process A then the two process owners need to sort out the reasons for their differing perceptions

(performance through intelligent management) and Justise (joined up systems thinking for integrating synergy in engineering).

5. THE MATHEMATICS OF ITALIAN FLAGS

An Italian flag is defined mathematically as an interval probability—so the probability of an event or proposition E has a lower bound El and an upper bound Eu. So $p(E)$ is an interval number as follows

$$p(E) = [El, Eu] = [g, (1 - r)]$$

Here g represents the green part of the flag colouring an interval on the scale [0, 1] from 0 to g and r is the red part of the flag from $(1-r)$ to 1. The white part of the flag is therefore $w = 1-g-r$.

Now consideration is given to the problem of calculating the Italian flag for a proposition H based on the flag for a single piece of evidence E. H is the proposition 'process H will be successful' and E is a sub-process of H at the next layer down in the PPM. The term $p(H)$ is therefore a measure of the dependability that the process H will be successful.

To calculate this, the total probability theorem is used

$$p(H) = p(H/E)\,p(E) + (1 - p(-H/-E))\,p(-E)$$

The values of $p(E)$ and $p(-E)$ are data from measurements, judgements or calculations. They may also be propagated from sub-processes. The values $p(H/E)$ and $p(-H/-E)$ are input by the user. The first measure $p(H/E)$ represents a warrant or justification for an action or a belief in the success of H if E is totally successful—it is the degree to which success in E provides positive support (assurance, affirmation) for the success of H. In a similar way $p(-H/-E)$ represents a warrant or justification for an action or a belief in the failure of H if E totally fails: it is the degree to which the failure of E provides

negative support (testability and falsification) for the failure of H.

So $p(H/E)$ is called the positive support and $p(-H/-E)$ is called the negative support. They are judged by asking the following questions for each sub-process separately regardless of all other sub-processes.

(a) Positive support: if E succeeds what are the chances that H will succeed regardless of the evidence for other sub-processes?

(b) Negative support: if E fails what are the chances that H will also fail regardless of the evidence for other sub-processes?

Figure 3 gives guidance. Note that $p(H/-E) = 1-p(-H/-E)$.

Now

$$p(H) = [Hl, \ Hu] \text{ and } p(H/E) = [Sn(H/E),$$
$$Sp(H/E)] = s = [sl, \ su]$$

where

$$Sn(H/E) = sl, \ Sp(H/E) = su$$

Likewise if the probability of not H is $p(-H)$ then

$$p(-H/-E) = [Sn(-H/-E),$$
$$Sp(-H/-E)] = n = [nl, \ nu]$$

so

$$Sn(-H/-E) = nl, \ Sp(-H/-E) = nu$$

Using probability theory it can be seen that

$$p(-H/E) = 1 - s = [Sn(-H/E),$$
$$Sp(-H/E)] = [1 - su, \ 1 - sl]$$

and

$$p(H/ - E) = 1 - n = [Sn(H/-E),$$
$$Sp(H/-E)] = [1 - nu, \ 1 - nl]$$

Calculate $p(H)$ using a variation on the total probability theorem as derived by Marashi[2] as follows

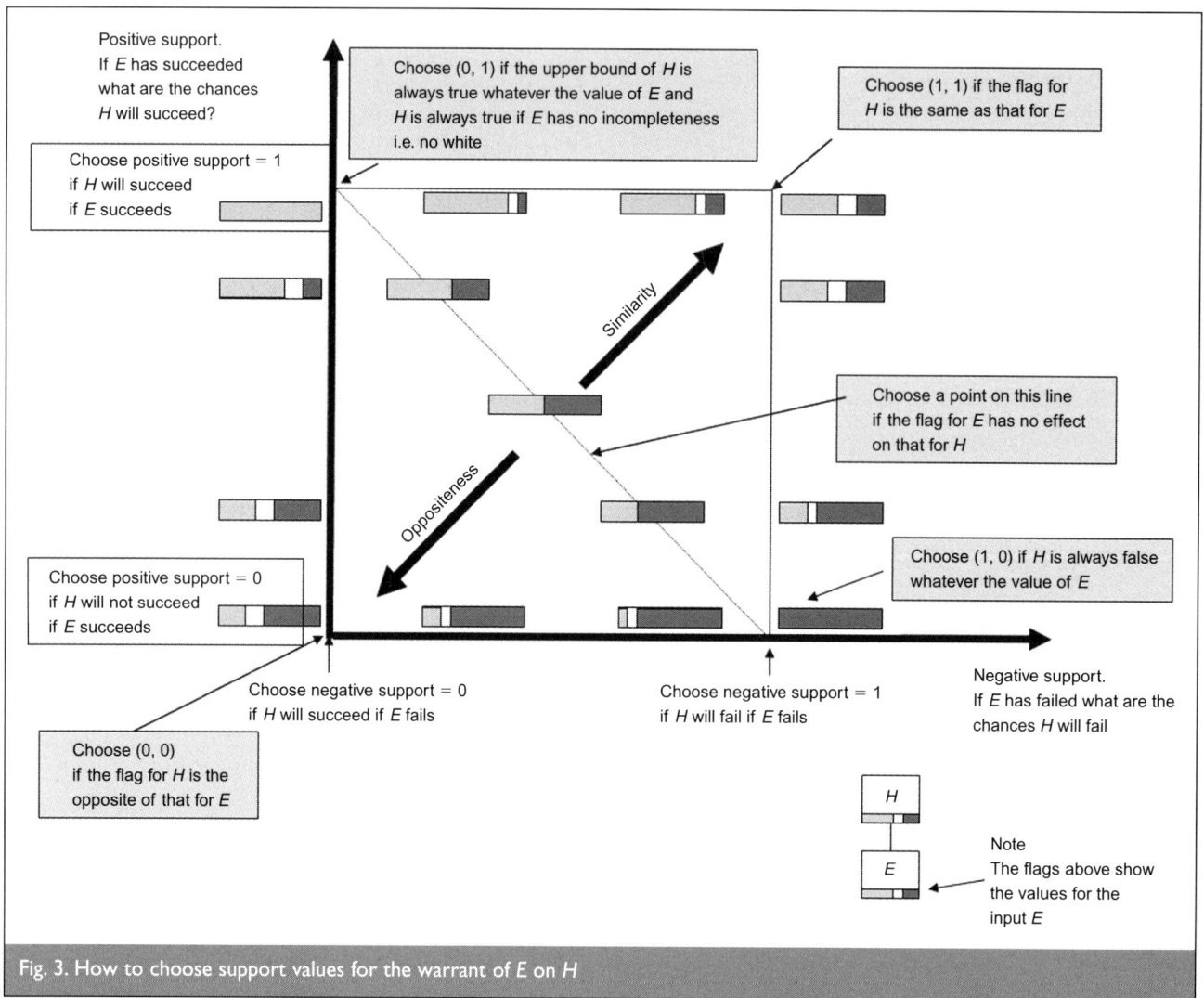

Fig. 3. How to choose support values for the warrant of E on H

1 $p(H) = p(H/E)g + p(H/-E)r + (p(H/E)\,p(H/-E))w$

2 $p(-H) = p(-H/E)g + p(-H/-E)r + (p(-H/E)\,p(-H/-E))w$

A simple interval version of the total probability theorem cannot be used because some of the basic probability constraints would be violated, as illustrated by the example below. To avoid these inconsistencies it is necessary to operate separately on the three intervals: g, w and r. The conditional probabilities operate on the g and r, but the white w is assumed to be a quantity independent of both conditionals, that is with least bias.

To find the upper bound on H the upper bound on $p(H/E)$ is used, that is, su operating on g and the upper bound on $p(H/-E)$, that is $(1-nl)$ operating on r. Thus using equation (1)

$Hu = Sp(H) = su\,g + (1 - nl)r + (1 - (1 - su)nl)\,w$

To calculate Hl, do the same but use the other bounds on s and n, that is sl and nu to calculate $-H$ before obtaining the

bounds on H. Thus the maximum of $-H$, that is $-Hu$, is calculated and then subtracted from 1 using equation (2) and

$$Hl = Sl(H) = 1 - Sp(-H)$$
$$= 1 - \{(1 - sl)g + nu\,r + (1 - sl(1 - nu))w\}$$

Here the upper bound is used on $p(-H/E)$ or $(1-s)$, that is $(1-sl)$, to maximise the green on $-H$ in order to minimise the green on H. The upper bound is used on $p(-H/-E)$ or nu to minimise the red on $-H$ and hence maximise the red on H.

In summary

$p(H) = [Sn(H), Sp(H)] = [Hl,\ Hu]$

where

$Hl = 1 - \{(1 - sl)g + nu\,r + (1 - sl(1 - nu))w\}$

$Hu = su\,g + (1 - nl)r + (1 - (1 - su)nl\,w$

Thus as an example
If $p(E) = [0\cdot3, 0\cdot7]$, that is $g = 0\cdot3$, $r = 0\cdot3$, $w = 0\cdot4$

$$[sl, su] = [0{\cdot}4, 0{\cdot}6]$$

$$[nl, nu] = [0{\cdot}4, 0{\cdot}8]$$

A simple interval calculus produces inconsistent results because

$$p(H) = [0{\cdot}4, \ 0{\cdot}6] \times [0{\cdot}3, \ 0{\cdot}7] + [1 - 0{\cdot}8, \ 1 - 0{\cdot}4]$$
$$\times [0{\cdot}3, \ 0{\cdot}7] = [0{\cdot}18, \ 0{\cdot}84]$$

and hence $p(-H) = [1 - 0{\cdot}84, 1 - 0{\cdot}18] = [0{\cdot}16, 0{\cdot}82]$

Whereas alternatively $p(-H) = [0{\cdot}4, 0{\cdot}6] \times [0{\cdot}3, 0{\cdot}7] + [0{\cdot}4, 0{\cdot}8] \times [0{\cdot}3, 0{\cdot}7] = [0{\cdot}24, 0{\cdot}98]$ and hence $p(H) = [0{\cdot}02, 0{\cdot}76]$.

These inconsistencies are removed using the formulae above so that

$$Hl = 1 - \{(1 - 0{\cdot}4) \times 0{\cdot}3 + 0{\cdot}8 \times 0{\cdot}3$$
$$+ \ (1 - 0{\cdot}4 \times (1 - 0{\cdot}8)) \times 0{\cdot}4$$
$$= 1 - \{0{\cdot}788\} = 0{\cdot}212$$

$$Hu = 0{\cdot}6 \times 0{\cdot}3 + (1 - 0{\cdot}4) \times 0{\cdot}3$$
$$+ \ (1 - (1 - 0{\cdot}6) \times 0{\cdot}4) \times 0{\cdot}4$$
$$= 0{\cdot}18 + 0{\cdot}18 + (1 - 0{\cdot}16) \times 0{\cdot}4$$
$$= 0{\cdot}36 + 0{\cdot}84 \times 0{\cdot}4 = 0{\cdot}696$$

and so

$$p(H) = [0{\cdot}212, 0{\cdot}696]$$

6. PROPAGATION OF EVIDENCE BY PAIRWISE COMBINATION

Now calculate the Italian flag of evidence $p(H)$ for a process H, which has sub-processes $E1$, $E2$, $E3$ and so on with associated supporting evidence flags of $p(E1)$, $p(E2)$, $p(E3)$ and so on.

A sub-process may be declared as jointly sufficient or jointly necessary. If a sub-process is declared as being jointly sufficient then the flag for any process calculated using that sub-process will not have a lower bound less than the lower bound for this sub-process. This means that, in this condition, the green must be at least as big as the green for the sub-process. Likewise if a sub-process is declared as being jointly necessary then the flag for any process calculated using that sub-process will not have an upper bound higher than the upper bound for that sub-process. This means that, in this condition, the red must be at least as big as the red for the sub-process.

The pairwise combination method begins by first calculating the evidence flag for a process based on the flag for each sub-process flag separately using the total probability theorem as described above. Any requirements of joint sufficiency or necessity are not included at this stage.

This therefore results in a series of separate estimates of $p(H)$ based on Ei, referred to as $p1(H)$, $p2(H)$, $p3(H)$.

These separate estimates are then combined in pairs. Dependency values λ need to be input for each pair combination.[2] The dependency values are a measure of the overlap or commonality between processes and are used to reduce the likelihood of 'double counting'. The dependency values range between total dependency, through independence to mutual exclusion and maximum perversity.

The first modelling assumption as set out in Fig. 4 is made by assuming that the dependency between $pi(H)$ and $pj(H)$ is as between $p(Ei)$ and $p(Ej)$.

If there is some support for H and also for $-H$ then this is conflicting evidence. This is automatically redistributed by the algorithm to the incomplete white range. Any conditions of joint sufficiency or necessity are included by checking the sizes of the green and red. This may result in inadmissible conflict which occurs when inputs are logically inconsistent. When this happens a blue section appears in the Italian flag, indicating its relative size. The only way to resolve this is by changing the input values.

A series of results $pij(H)$ are now available and they must be combined into a single answer. This leads to a second modelling assumption with two alternatives suitable for different purposes. The first is to assume that the results $pij(H)$ are all totally dependent. Here the minimum green (smallest positive evidence) and minimum red (smallest negative evidence) are taken and this produces maximum white (maximum don't know). The second is that average green and the average red are taken to give an indication of the total effect of the positive and negative support for H. Any requirements of joint sufficiency or necessity are included at this stage.

The choice of which way to combine the $pij(H)$ will be made according to the needs of the process players. It is important to remember that the purpose of this methodology is not to provide accurate predictions of the future—rather it is to enable the players to understand the sources of uncertainty and hence to make appropriate decisions to improve the chances of attaining success. Both of these assumptions can reveal different particular insights.

7. A STRUCTURAL EXAMPLE: PROCURING A NEW BUILDING

Assume that the main area of interest is the risks to the procurement of a new building. Although structural engineers may often not be concerned with such a high-level process, it is a critical context for the success of their work. Following the process methodology described by Blockley and Godfrey[1] the first task is to identify a process model, as shown in Fig. 5. Note that, owing to space restrictions, the model shown in the figure is incomplete and it is not possible to describe the methodology in any great detail. The first point to note is that the model is hierarchical and so will include the totality of the design as a sub-process, which will in turn include the structural design. The process is therefore 'procuring a new building' and represents the whole of the project recognising the complexity of all of its

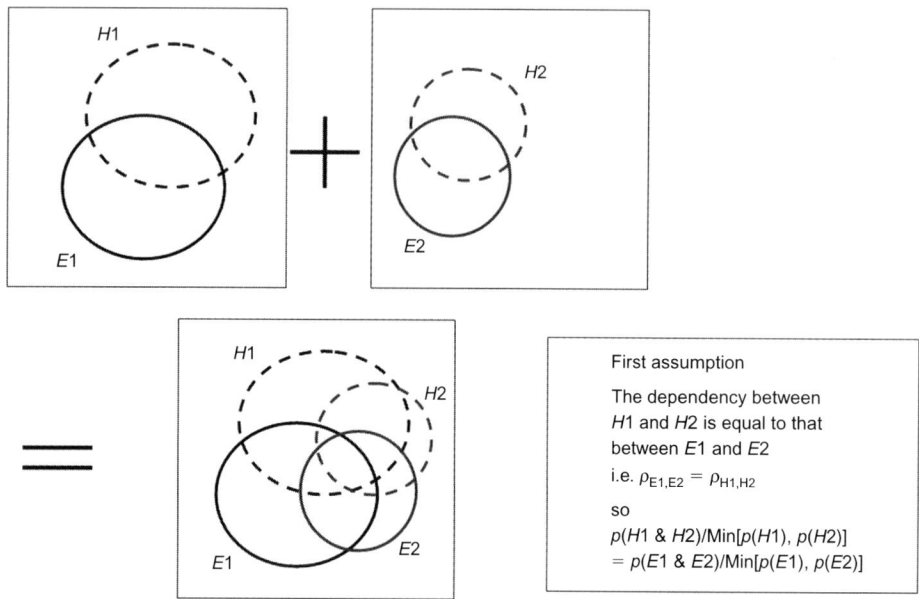

Fig. 4. Nature of first assumption

First assumption

The dependency between $H1$ and $H2$ is equal to that between $E1$ and $E2$

i.e. $\rho_{E1,E2} = \rho_{H1,H2}$

so

$p(H1 \text{ \& } H2)/\text{Min}[p(H1), p(H2)]$
$= p(E1 \text{ \& } E2)/\text{Min}[p(E1), p(E2)]$

emergent properties. Its associated Italian flag is therefore an indicator of the total risk to the success of the whole project and will change in time as evidence about the risks change. Formally the flag illustrates an interval probability measure of the dependability that the available evidence indicates that the process will be successful. Of course that means that success has to be defined precisely. Before the start of the project there is no evidence at all about the risks and so the flag will be completely white.

The model is developed by identifying at the second level down all of the sub-processes

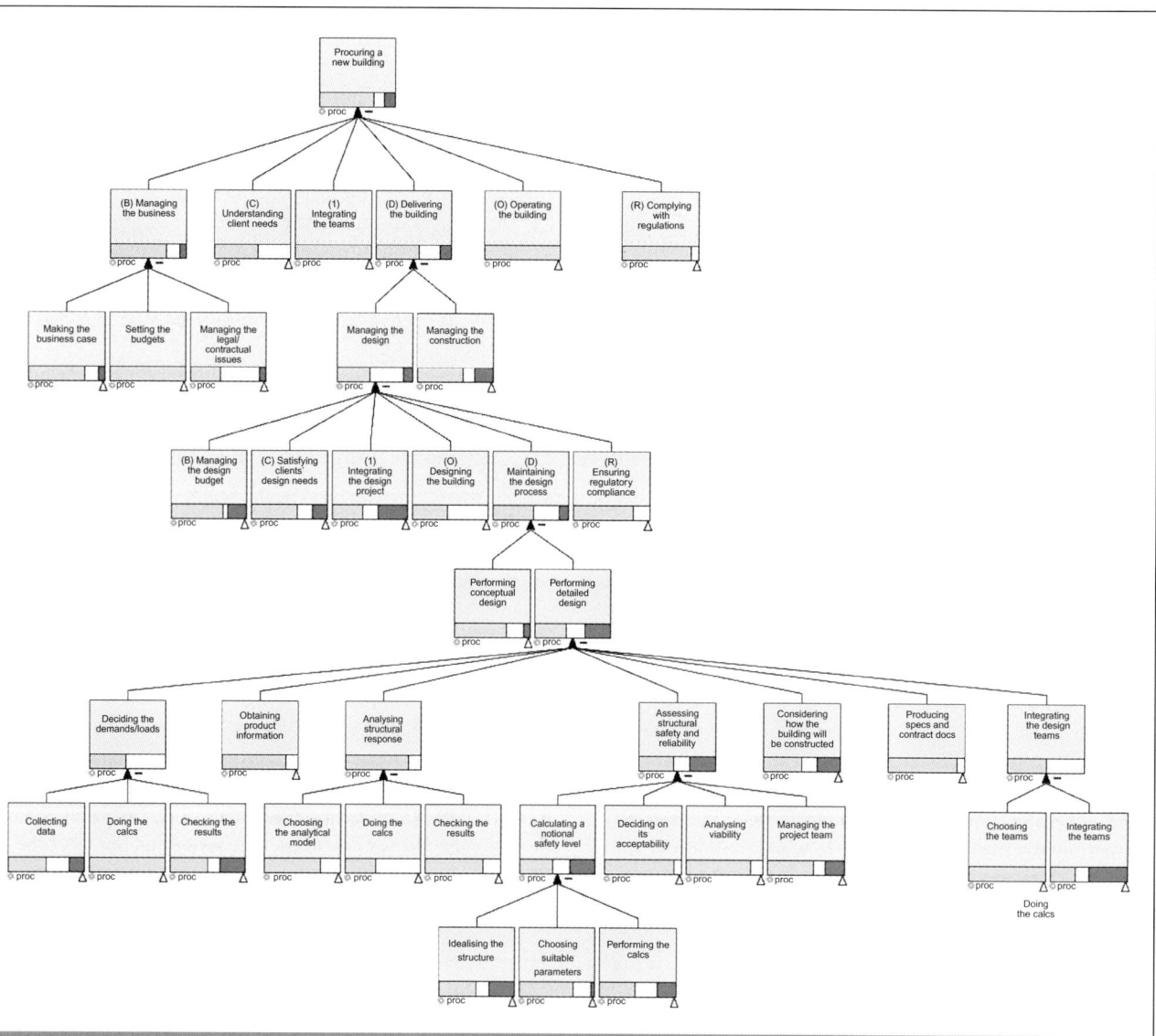

Fig 5. Example process model—procuring a new building (part)

which together are necessary and sufficient for the top process to succeed. Each has its own Italian flag, which at the start are also white. One of the six sub-processes in Fig. 5 is 'delivering the building.' Then by looking at each second layer process in turn the third level processes are identified. Again they are the processes, the successes of which are together necessary and sufficient for the success of a specific chosen second level process. 'Delivering the building' has 'managing the design' as one of its six sub-processes. Again each has an Italian flag. The building of the model is repeated for further levels to an appropriate level of detail such as 'performing the calculations' in Fig. 5.

The Italian flags for the bottom processes in particular (or all processes in general) can either be input directly by the relevant process owner or derived from a performance indicator using a value function as described by Hall et al.[3] The flags for the higher process, right to the very top, are then calculated using the pair wise algorithm described earlier so that the effects of the lower flags on the higher flags become evident. At the same time, process owners will have their own interpretations of the evidence available to them and hence their own versions of the flags. Any important discrepancies as illustrated by Fig. 2 are highlighted and appropriate decisions taken in a timely manner.

The system is dynamic and constantly changing as new evidence arises and process owners change their flags. The system is also a rich source of project information as each process has information attached to it (some of it emergent) under the labels 'who, what, where, when and how'. Perhaps the most important attribute is that each process has an 'owner' who monitors the flag and is responsible for identifying actions required to put the process back on track. This, of course, may well involve negotiating with others about changes they may need to make, since processes are interdependent.

Note that Fig. 5 includes all processes—both soft and hard. In other words the model is not restricted to the physical building but includes all information of all kinds. The pedigree of the information clearly varies enormously and that is catered for through the input data. Thus if a difficult judgement, which is necessarily imprecise, is put alongside information from a precise measurement or calculation then this can be allowed for by a suitable adjustment of the relationships between the measurement and the flag. The computer systems Justise also allows notes to be written and links to be made to any relevant documents so that process owners and players as well as stakeholders with access permission can immediately see any changes.

So how is the model used and has it any relevance to a practising structural engineer? If implemented on a project intranet the model would be available to all participants in a project (with appropriate security safeguards) with read only or write access to their 'pieces of the action'. Process owners would be responsible for feeding in their latest data. After an initial input of contact details and other relatively stable information the major changes may be values of state variables (e.g. hard systems: movement of a foundation or dynamic responses of a bridge deck), key performance indicators (KPIs) (e.g. soft systems–judgements about team behaviour or loss of key personnel) and/or judgements about the Italian flags. The inputs can include the results of very detailed calculations such as finite element analysis or even notional probabilities of failure. Other examples are detailed design and construction activities together with judgements and notes about the context and dependability of any calculations. In this way they can be interpreted and integrated into the total picture as dependably as possible.

A quick scan over all the flags will enable the spotting of possibly previously hidden interactions that could cause problems. In this way, decisions are taken by everyone involved in full awareness of the impact of other decision makers on total project success.

8. CONCLUSIONS

(a) The need to integrate different risks from different types of information has been highlighted. It has been argued that structural engineers have to understand and deal with the risks to their projects from the complex interactions between technical and non-technical issues.

(b) The use of process maps to connect data, information, people and functions and provide routes by which change can be managed has been described and illustrated with an example of the procuring of a building.

(c) The clear distinction between hard and soft systems has been made. It has been shown that they can be integrated by modelling them both as an interacting set of process holons.

(d) Italian flags based on interval probability theory are a practical way of representing the dependability of evidence in a process model. Some new mathematics of interval probability has been presented, which enables the interval values to be propagated up through the hierarchy of a process model.

(e) The Italian flag can be used as part of a social process to improve understanding of, and judgements associated with, incompleteness of knowledge.

REFERENCES

1. BLOCKLEY D. I. and GODFREY P. S. *Doing it Differently*. Thomas Telford, London, 2000.
2. MARASHI E. *Managing Discourse and Uncertainty for Decision Making in Civil and Infrastructure Engineering Systems*. PhD thesis, University of Bristol, UK, 2006.
3. HALL J. W., LE MASURIER J. W., BAKER-LANGMAN E. J., DAVIS J. P. and TAYLOR C. A. A decision-support methodology for performance-based asset management. *Civil Engineering and Environmental Systems*, 2004, 21, No. 1, 51–75.

What do you think?

To comment on this paper, please email up to 500 words to the editor at journals@ice.org.uk

Proceedings journals rely entirely on contributions sent in by civil engineers and related professionals, academics and students. Papers should be 2000–5000 words long, with adequate illustrations and references. Please visit www.thomastelford.com/journals for author guidelines and further details.